LAMBDA-RINGS

LAMBDA-RINGS

Donald Yau

The Ohio State University at Newark, USA

NEW JERSEY • LONDON • SINGAPORE • BEIJING • SHANGHAI • HONG KONG • TAIPEI • CHENNAI

Published by

World Scientific Publishing Co. Pte. Ltd.
5 Toh Tuck Link, Singapore 596224
USA office: 27 Warren Street, Suite 401-402, Hackensack, NJ 07601
UK office: 57 Shelton Street, Covent Garden, London WC2H 9HE

British Library Cataloguing-in-Publication Data
A catalogue record for this book is available from the British Library.

LAMBDA-RINGS

ISBN-13 978-981-4299-09-1
ISBN-10 981-4299-09-X

Printed in Singapore

To Eun Soo and Hye-Min

Preface

Some History

A λ-*ring* is a commutative ring R with unit together with operations

$$\lambda^i : R \to R \quad (i \geq 0)$$

that behave like exterior power operations. It was originally introduced by Berthelot [Berthelot (1971)] and Grothendieck [Grothendieck (1971)] in the context of K-theory to study the Grothendieck-Riemann-Roch Theorem. In the setting of K-theory, the λ-operation λ^i is induced by the ith exterior power on vector bundles.

Among the mathematical fields in which λ-rings play a role are:

- K-theory (topological K-theory, algebraic K-theory, Grothendieck-Riemann-Roch theory, and Hochschild and cyclic homology of commutative algebras).
- Representation theory (representation rings of connected compact Lie groups, classification of connected compact Lie groups through their classifying spaces, and Burnside rings of finite groups).
- Witt vectors type constructions (formal group laws, algebraic geometry, number theory, generalized cohomology theories, combinatorics of necklace rings, and the cyclotomic identity),
- Binomial rings and integer-valued polynomials.
- Convex polytopes (polytope groups and toric varieties).
- Class field theory (finite étale **Q**-algebras, cyclotomic fields, and (un)ramified primes).

Here we mention a few relevant results to illustrate the power of λ-rings.

(1) One such example is the very simple proof by Adams and Atiyah [Adams and Atiyah (1966)] of Adams's theorem [Adams (1960)] about the non-existence of elements of Hopf invariant one in the homotopy group $\pi_{2n-1}(S^n)$ when $n \neq 1, 2, 4$, or 8. The original proof by Adams [Adams (1960)] was fairly long and used secondary operations in ordinary cohomology. Using K-theory and Adams

operations, which exist on any λ-ring, Adams and Atiyah were able to prove this theorem in under two pages.

(2) Another use of λ-rings in homotopy theory is the classification of the Mislin genus of the classifying space of a Lie group. Given a 1-connected compact Lie group G, it is known that the Mislin genus of its classifying space BG is uncountable [Møller (1992); Rector (1971)]. In other words, there exist uncountably many spaces X of distinct homotopy types such that the p-localizations $X_{(p)}$ and $BG_{(p)}$ are homotopy equivalent at each prime p and also the rationalizations $X_{(0)}$ and $BG_{(0)}$ are homotopy equivalent. In [Notbohm (1993)] Notbohm showed that the Mislin genus of BG is classified by K-theory λ-rings. So any two distinct homotopy types in the Mislin genus of BG can be distinguished by their K-theory λ-rings.

(3) In representation theory, the complex representation ring $R(G)$ of a group G is a λ-ring, in which λ^i is induced by the ith exterior power on representations of G. When G is a connected compact Lie group, a major result of Osse [Osse (1997)] states that G is uniquely determined by the λ-ring $R(G)$, up to a certain product of $Sp(k)$ or $SO(2k+1)$. Osse then used this result to show that G is uniquely determined by its classifying space BG. Moreover, it is possible to describe the λ-ring $R(G)$ completely [Guillot (2007)].

(4) The ring $\mathbf{W}(A)$ of big Witt vectors of a ring A plays an important role in the theory of formal groups [Hazewinkel (1978)]. The ring of big Witt vectors is actually a λ-ring. In fact, the ring $\mathbf{W}(A)$ is canonically isomorphic to the universal λ-ring $\Lambda(A)$ via the Artin-Hasse Exponential isomorphism. Furthermore, when A is a binomial ring, $\Lambda(A)$ is isomorphic to the necklace ring $Nr(A)$ of Metropolis and Rota [Metropolis and Rota (1983)]. The construction of the necklace ring is motivated by certain combinatorial identities about the number of primitive necklaces.

Besides the original papers by Berthelot [Berthelot (1971)] and Grothendieck [Grothendieck (1971)], there are three classic references for λ-rings. The first one and the most readable account of λ-rings is the paper [Atiyah and Tall (1969)] by Atiyah and Tall, which applies λ-rings to study group representations and equivariant K-theory. The second one is Knutson's lecture notes [Knutson (1973)], which applies λ-rings to study representations of the symmetric groups. The third one is the book [Fulton and Lang (1985)] by Fulton and Lang, which is about the Grothendieck-Riemann-Roch theory in algebraic geometry. There is also a survey article [Patras (2003)] by Patras that outlines some of the applications of λ-rings and contains a long list of references related to λ-rings.

Purposes of This Book

In addition to being powerful tools in several fields of mathematics, λ-rings are interesting algebraic objects in their own right. One main theme of this book is that λ-rings can and should be studied purely algebraically. There are interesting purely algebraic results about λ-rings, and there are related open questions yet to be answered.

For example, it turns out that many rings of interest, such as power series rings and most of their quotients, admit a lot of λ-ring structures. In fact, the following quotient power series ring

$$R = \frac{\mathbf{Z}[[x_1, \ldots, x_n]]}{\langle x_1^{e_1}, \ldots, x_n^{e_n} \rangle}$$

admits *uncountably many isomorphism classes* of filtered λ-ring structures, where $n \geq 1$ and $2 \leq e_i \leq \infty$. (A filtered λ-ring is a λ-ring together with a compatible filtration.) This statement includes the biggest cases of the power series rings $\mathbf{Z}[[x_1, \ldots, x_n]]$ in n variables (when each $e_i = \infty$) and the smallest case of the dual number ring $\mathbf{Z}[x]/\langle x^2 \rangle$.

The main purpose of this book is to give an elementary introduction to λ-rings. With the exceptions of a few examples, Corollaries, and Chapter 8, the whole book is purely algebraic. The materials are aimed at beginning graduate students. In fact, one year of undergraduate abstract algebra is sufficient to understand most of this book. With this purpose in mind, this book is written with a lot more details than the sources cited above. The author also hopes that this book will serve as a general reference for researchers who use λ-rings and closely related objects in their work, be that homotopy theory, algebraic geometry, representation theory, or other fields.

There are eight chapters in this book. The first seven chapters contain the core materials. The last chapter contains a list of open problems related to λ-rings. Chapters 1 and 2 and part of Chapter 3 are mostly based on [Atiyah and Tall (1969); Knutson (1973)], but we organize the materials differently to improve readability. Chapter 4 on Witt vectors mostly follows Chapter III in [Hazewinkel (1978)]. Chapter 5 on binomial rings is based on [Elliott (2006); Wilkerson (1982)]. Chapters 6 and 7 follow the author's papers [Yau (2004a, 2006)].

Several examples and Corollaries deal with topological K-theory and representation rings. For introductory materials on K-theory, the reader may consult [Atiyah (1989); Kono and Tamaki (2002); May (1999)]. For basic materials on representations, the reader may consult [Fulton and Harris (1991); James and Liebeck (2001)].

The Chapters

Below we give a brief description of each chapter in this book.

Chapter 1 gives the definition of λ-rings and constructs the free λ-ring on one generator. Then we prove the Verification Principle, which states that a natural operation on λ-rings is uniquely determined by its action on sums of one-dimensional elements. The last section of this chapter contains the Splitting Principle, which is the algebraic counterpart of a result of the same name in K-theory. It states that, up to an extension of λ-rings, an element in a λ-ring of dimension n can be written as a sum of n one-dimensional elements.

In Chapter 2 we provide a more conceptual description of λ-rings. The comonad Λ is constructed, and λ-rings are exactly the Λ-coalgebras. Given a ring R, the λ-ring $\Lambda(R)$ is called the *universal λ-ring of R*.

In Chapter 3 we begin with the construction of the Adams operations ψ^n on a λ-ring. Unlike the λ-operations, which are neither additive nor multiplicative, the Adams operations ψ^n $(n \geq 1)$ are ring endomorphisms that also commute with all the λ-operations. Moreover, Adams operations act particularly nicely on the quotients of the γ-filtration, which exists on any augmented λ-ring. We prove an important result due to Wilkerson. It states that for a $\mathbf{Z}$-torsion free ring R, giving a λ-ring structure on R is equivalent to specifying ring endomorphisms ψ^n $(n \geq 1)$ on R that behave like the Adams operations. So a $\mathbf{Z}$-torsion free λ-ring is uniquely determined by its Adams operations. As applications we show that, under some mild assumptions, localizations and completions of λ-rings are again λ-rings.

In Chapter 4 we describe another incarnation of the universal λ-ring. Namely, $\Lambda(R)$ is isomorphic to the ring $\mathbf{W}(R)$ of *big Witt vectors on R*. Witt vectors are important objects in the theory of formal groups and combinatorics, among others. The big Witt vector construction forms a comonad $\mathbf{W}$, which is isomorphic to the comonad Λ via the Artin-Hasse Exponential isomorphism.

In Chapter 5 we discuss *binomial ring*, which is equivalent to a λ-ring in which all the Adams operations are equal to the identity map. Another characterization of a binomial ring, due to Elliott [Elliott (2006)], is that it is $\mathbf{Z}$-torsion-free and is the homomorphic image of a ring of integer-valued polynomials. Knutson's Theorem states that a binomial ring with a certain kind of generating subset must be isomorphic to the ring of integers $\mathbf{Z}$. When R is binomial, there is an isomorphism between the universal λ-ring $\Lambda(R)$ and $Nr(R)$, where $Nr(R)$ is the Metropolis-Rota [Metropolis and Rota (1983)] necklace ring of R.

Chapter 6 is devoted to the purely algebraic question of how many λ-ring structures a ring can possess. We introduce the notion of a *filtered λ-ring*, which is a λ-ring that comes equipped with a decreasing filtration of λ-ideals. For example, any augmented λ-ring is a filtered λ-ring when equipped with the γ-filtration. Consider the quotient power series or polynomial ring

$$R = \frac{A[[x_1, \ldots, x_n]]}{\langle x_1^{e_1}, \ldots, x_n^{e_n} \rangle} \quad \text{or} \quad \frac{A[x_1, \ldots, x_n]}{\langle x_1^{e_1}, \ldots, x_n^{e_n} \rangle},$$

in which A is any subring of $\mathbf{Q}$, $n \geq 1$, and at least one $e_j < \infty$ (where $x_i^\infty = 0$ by definition). We show that any such filtered ring R admits uncountably many isomorphism classes of filtered λ-ring structures. In the two smallest cases, $R = \mathbf{Z}[x]/\langle x^2 \rangle$ or $\mathbf{Z}[x]/\langle x^3 \rangle$, we provide a complete classification of all the filtered λ-ring structures on R.

In Chapter 7 we continue the study of the algebraic question of Chapter 6. It is shown that the function

$$A \mapsto \{\text{filtered } \lambda\text{-ring structures on } A[[x]]\}$$

for subrings $A \subseteq \mathbf{Q}$ is corepresentable and that the corepresenting object is a binomial ring. In the rest of Chapter 7, we extend the universal λ-ring functor Λ to the category of filtered rings. Filtered λ-rings are exactly the coalgebras of the resulting comonad.

In Chapter 8 we put together a list of open problems that are related to λ-rings. Some of these problems are about the purely algebraic properties of λ-rings. The other problems have to do with applications of λ-rings to other fields.

Donald Yau

Contents

Chapter 1

λ-Rings

The definition of a λ-ring involves certain polynomials that arise in the theory of symmetric functions. We will, therefore, begin by discussing symmetric functions in §1.1. This material is fairly standard and can be found in most books on abstract algebra, e.g., Chapter 2 in [Jacobson (1985)].

The definition of a λ-ring and some of its basic properties are given in §1.2. In §1.3 we discuss the free λ-ring on one generator. This is used in §1.4 to prove the Verification Principle, which states that a natural operation on λ-rings is uniquely determined by its actions on sums of one-dimensional elements. The Splitting Principle is discussed in §1.5. This result allows one to write an n-dimensional element in a λ-ring as a sum of n one-dimensional elements. The materials on the free λ-ring, the Verification Principle, and the Splitting Principle originated in the paper [Atiyah and Tall (1969)] by Atiyah and Tall. Another reference is Knutson's book [Knutson (1973)]. The original sources for λ-rings are [Berthelot (1971); Grothendieck (1971)].

Throughout this book, we use the following standard notations:

- $\mathbf{Z}$, the integers,
- $\mathbf{Z}_{(p)}$, the p-local integers for a prime p,
- $\mathbf{Q}$, the rationals,
- $\mathbf{C}$, the complex numbers.

Unless otherwise specified, by a *ring* we always mean an associative and commutative ring with a multiplicative identity $1 \neq 0$. Ring homomorphisms are required to preserve the multiplicative identity. An object R that satisfies all the ring axioms except for the existence of a multiplicative identity is called a *non-unital ring*. Non-unital rings will be used in §3.3.

1.1 Symmetric Functions

Fix a ring R. Let $R[x_1, \ldots, x_n]$ be the polynomial ring over R in n independent variables $x_1, \ldots, x_n$. If $f \in R[x_1, \ldots, x_n]$ and π is a permutation on $\{1, \ldots, n\}$, then

πf is defined as the polynomial

$$\pi f = f(x_{\pi(1)}, \ldots, x_{\pi(n)}).$$

A polynomial $f \in R[x_1, \ldots, x_n]$ is called a *symmetric function* if $\pi f = f$ for every permutation π on $\{1, \ldots, n\}$.

Example 1.1. For $1 \leq k \leq n$, let $s_k \in R[x_1, \ldots, x_n]$ be the polynomial

$$s_k = \sum_{1 \leq i_1 < \cdots < i_k \leq n} x_{i_1} x_{i_2} \cdots x_{i_k}.$$

For example, we have

$$\begin{aligned} s_1 &= x_1 + \cdots + x_n, \\ s_2 &= x_1 x_2 + x_1 x_3 + \cdots + x_{n-1} x_n, \\ s_n &= x_1 x_2 \cdots x_n. \end{aligned}$$

Each s_k is a symmetric function, since s_k is the coefficient of t^{n-k} in

$$\begin{aligned} \prod_{k=1}^{n} (t - x_k) &= (t - x_1)(t - x_2) \cdots (t - x_n) \\ &= t^n - s_1 t^{n-1} + s_2 t^{n-2} - \cdots + (-1)^n s_n. \end{aligned} \tag{1.1}$$

The polynomial s_k is called the *kth elementary symmetric function on* $x_1, \ldots, x_n$. We will sometimes write $s_k(x_1, \ldots, x_n)$ for s_k to avoid any ambiguity.

Another way to define the elementary symmetric functions is by looking at their generating function:

$$\prod_{k=1}^{n} (1 + x_i t) = 1 + s_1 t + s_2 t^2 + \cdots + s_n t^n. \tag{1.2}$$

Note that

$$s_k(x_1, \ldots, x_n, 0) = s_k(x_1, \ldots, x_n).$$

Example 1.2. For any positive integer r, the rth power sum

$$x_1^r + \cdots + x_n^r$$

is a symmetric function.

Example 1.3. For any positive integers m and n, consider the polynomial

$$g(t) = \prod_{1 \leq i_1 < \cdots < i_m \leq mn} (1 + x_{i_1} \cdots x_{i_m} t)$$

Then the coefficient of each t^j in g is a symmetric function in the variables $x_1, \ldots, x_{mn}$.

A fundamental property of symmetric functions is that they are all generated by the elementary symmetric functions.

Theorem 1.4 (The Fundamental Theorem of Symmetric Functions). *The elementary symmetric functions $s_1, \ldots, s_n$ in $R[x_1, \ldots, x_n]$ are algebraically independent over R. Moreover, every symmetric function f in $R[x_1, \ldots, x_n]$ can be written uniquely as a polynomial with coefficients in R in the elementary symmetric functions.*

Before giving the proof, we need the following terminology. The *degree* of a monomial

$$x_1^{k_1} \cdots x_n^{k_n} \tag{1.3}$$

is $k_1 + \cdots + k_n$. A polynomial $f \in R[x_1, \ldots, x_n]$ is said to be *homogeneous of degree m* if every non-zero monomial in f has degree m. We order monomials *lexicographically*. In other words, we say that

$$x_1^{k_1} \cdots x_n^{k_n} \quad \text{is higher than} \quad x_1^{l_1} \cdots x_n^{l_n} \tag{1.4}$$

if there exists $s \in \{1, \ldots, n\}$ such that $k_i = l_i$ for $i < s$ and $k_s > l_s$.

Example 1.5. The elementary symmetric function $s_k = s_k(x_1, \ldots, x_n)$ is homogeneous of degree k. The highest monomial in s_k is $x_1 x_2 \cdots x_k$. In general, given non-negative integers $l_1, \ldots, l_n$, the polynomial $s_1^{l_1} \cdots s_n^{l_n}$ is homogeneous of degree

$$l_1 + 2l_2 + \cdots + nl_n = \sum_{k=1}^{n} kl_k.$$

Its highest monomial is

$$x_1^{l_1+\cdots+l_n} x_2^{l_2+\cdots+l_n} \cdots x_n^{l_n}.$$

Moreover, if $s_1^{j_1} \cdots s_n^{j_n}$ has the same highest monomial as that of $s_1^{l_1} \cdots s_n^{l_n}$, then $j_n = l_n$ from the exponents of x_n. Therefore, $j_{n-1} = l_{n-1}$ from the exponents of x_{n-1}, and so forth. So $j_i = l_i$ for $i = 1, \ldots, n$.

Proof. [Proof of Theorem 1.4] First we prove that every symmetric function f in $R[x_1, \ldots, x_n]$ can be written as a polynomial with coefficients in R in the elementary symmetric functions. It suffices to prove this when f is homogenous, say, of degree m. Suppose that $x_1^{k_1} \cdots x_n^{k_n}$ is the highest monomial in f with non-zero coefficient. Since f is symmetric, it has a monomial of the form $x_{\pi(1)}^{k_1} \cdots x_{\pi(n)}^{k_n}$ for any permutation π. Since $x_1^{k_1} \cdots x_n^{k_n}$ is the highest monomial in f, it follows that $k_i \geq k_{i+1}$ for $i = 1, \ldots, n-1$.

From Example 1.5, the polynomial $s_1^{k_1-k_2} s_2^{k_2-k_3} \cdots s_n^{k_n}$ is homogeneous of degree $k_1 + \cdots + k_n = m$ and has highest monomial $x_1^{k_1} \cdots x_n^{k_n}$. If the coefficient of $x_1^{k_1} \cdots x_n^{k_n}$ in f is r, then the polynomial

$$f_1 = f - rs_1^{k_1-k_2} s_2^{k_2-k_3} \cdots s_n^{k_n}$$

is homogeneous of degree m, whose highest monomial is less than $x_1^{k_1} \cdots x_n^{k_n}$. Now we apply the same procedure to f_1, and so forth. The procedure has to stop after a finite number of steps, since there are only finitely many monomials in $R[x_1, \ldots, x_n]$ of degree m. This shows that every symmetric function $f \in R[x_1, \ldots, x_n]$ can be written as a polynomial in the elementary symmetric functions.

Next we show that the elementary symmetric functions are algebraically independent over R. Indeed, if they are not algebraically independent, then there is a non-trivial relation

$$\sum r_{(d_1,\ldots,d_n)} s_1^{d_1} \cdots s_n^{d_n} = 0$$

in which every coefficient $r_{(d_1,\ldots,d_n)} \neq 0$ in R. Among the multi-indexes $(d_1, \ldots, d_n)$, choose the one for which $s_1^{d_1} \cdots s_n^{d_n}$ has the highest degree and the highest monomial, say, $x_1^{k_1} \cdots x_n^{k_n}$. From Example 1.5 again, this monomial only appears once. This is a contradiction to the algebraic independence of the variables $x_1, \ldots, x_n$. This proves that the elementary symmetric functions are algebraically independent.

The uniqueness assertion in the Theorem now follows from the algebraic independence of the elementary symmetric functions. □

Example 1.6. The square sum can be expressed as

$$x_1^2 + \cdots + x_n^2 = (x_1 + \cdots + x_n)^2 - 2\left(\sum_{i<j} x_i x_j\right) = s_1^2 - 2s_2.$$

In general, there is a polynomial Q_r with integer coefficients such that

$$x_1^r + \cdots + x_n^r = Q_r(s_1, \ldots, s_n).$$

These polynomials Q_r will be used in Theorem 3.9 to express Adams operations in terms of λ-operations.

Example 1.7. In the polynomial

$$\prod_{1 \leq i < j \leq 4} (1 + x_i x_j t),$$

the coefficient of t is s_2 and the coefficient of t^6 is s_4^3. In general, there is a universal polynomial $P_{n,m}$ in nm variables with integer coefficients such that the coefficient of t^n in

$$g(t) = \prod_{1 \leq i_1 < \cdots < i_m \leq nm} (1 + x_{i_1} \cdots x_{i_m} t)$$

is $P_{n,m}(s_1, \ldots, s_{nm})$. These polynomials $P_{n,m}$ will appear in the definition of a λ-ring (Definition 1.10).

For example, $P_{n,1}(s_1, \ldots, s_n)$ is the coefficient of t^n in

$$\prod_{1 \leq i \leq n} (1 + x_i t) = (1 + x_1 t) \cdots (1 + x_n t),$$

which is $x_1 \cdots x_n = s_n$. In other words, we have

$$P_{n,1}(s_1, \ldots, s_n) = s_n. \tag{1.5}$$

Similarly, $P_{1,m}(s_1, \ldots, s_m)$ is the coefficient of t in

$$\prod_{1 \leq i_1 < \cdots < i_m \leq m} (1 + x_{i_1} \cdots x_{i_m} t) = 1 + (x_1 \cdots x_m)t.$$

So we have

$$P_{1,m}(s_1, \ldots, s_m) = s_m. \tag{1.6}$$

1.1.1 *Symmetric functions in two sets of variables*

We will need a slight generalization of Theorem 1.4. Let $y_1, \ldots, y_n$ be another set of variables, and let $\sigma_1, \ldots, \sigma_n$ be their elementary symmetric functions. Let f be an element in the polynomial ring $R[x_1, \ldots, x_n; y_1, \ldots, y_n]$. We say that f is a *symmetric function* if

$$f(x_1, \ldots, x_n; y_1, \ldots, y_n) = f(x_{\pi(1)}, \ldots, x_{\pi(n)}; y_{\tau(1)}, \ldots, y_{\tau(n)})$$

for every pair of permutations π and τ on $\{1, \ldots, n\}$.

Theorem 1.8. *Every symmetric function f in $R[x_1, \ldots, x_n; y_1, \ldots, y_n]$ can be written uniquely as a polynomial with coefficients in R in the elementary symmetric functions $s_1, \ldots, s_n$ and $\sigma_1, \ldots, \sigma_n$.*

Proof. Use Theorem 1.4 twice. The details are left to the reader as an easy exercise. □

Example 1.9. The coefficient of each t^k in

$$h(t) = \prod_{i,j=1}^{n} (1 + x_i y_j t)$$

is a symmetric function on the x_i's and the y_j's. Therefore, it must be a polynomial in the elementary symmetric functions $s_1, \ldots, s_n$ and $\sigma_1, \ldots, \sigma_n$. For example, a little bit of computation shows that

$$\prod_{i,j=1}^{2} (1 + x_i y_j t) = 1 + (s_1 \sigma_1)\, t + \left(s_1^2 \sigma_2 + s_2 \sigma_1^2 - 2 s_2 \sigma_2\right) t^2 \\ + (s_1 s_2 \sigma_1 \sigma_2)\, t^3 + \left(s_2^2 \sigma_2^2\right) t^4.$$

The reader should check this.

In general, the coefficient of t^n in $h(t) = \prod_{i,j=1}^{n}(1 + x_i y_j t)$ is of the form

$$P_n(s_1, \ldots, s_n; \sigma_1, \ldots, \sigma_n)$$

for some universal polynomial P_n in $2n$ variables with integer coefficients. For example, we have

$$P_1(x; y) = xy,$$
$$P_2(x_1, x_2; y_1, y_2) = x_1^2 y_2 + x_2 y_1^2 - 2 x_2 y_2.$$

These polynomials P_n will appear in the definition of a λ-ring (Definition 1.10).

1.2 λ-Rings

The definition of a λ-ring involves certain axioms about the operations λ^0, λ^1, $\lambda^n(x+y)$, $\lambda^n(xy)$, and $\lambda^m(\lambda^n(x))$. Before giving the exact definition, let us provide some motivation for the axioms.

1.2.1 *The λ-ring* Z

The simplest λ-ring is the ring of integers $\mathbf{Z}$, in which

$$\lambda^i(n) = \binom{n}{i} = \frac{n!}{i!(n-i)!}$$

for positive integers n. The corresponding generating function is

$$\lambda_t(n) = \sum_{i=0}^{\infty} \lambda^i(n)t^i = (1+t)^n,$$

which applies to all integers n. In other words, $\lambda^i(n)$ is the coefficient of t^i in $\lambda_t(n)$. Since

$$(1+t)^n = 1 + nt + \cdots,$$

we have $\lambda^0(n) = 1$, $\lambda^1(n) = n$, and $\lambda_t(1) = 1 + t$. Given two integers m and n, we have

$$\begin{aligned}\lambda_t(m+n) &= (1+t)^{m+n} \\ &= (1+t)^m(1+t)^n \\ &= \lambda_t(m)\lambda_t(n).\end{aligned}$$

So

$$\lambda^k(m+n) = \sum_{i+j=k} \lambda^i(m)\lambda^j(n),$$

which expresses λ^k of a sum in terms of the individual λ^i. Furthermore, a little bit of computation shows that

$$\begin{aligned}\binom{mn}{2} &= m^2\binom{n}{2} + n^2\binom{m}{2} - 2\binom{m}{2}\binom{n}{2}, \\ \binom{\binom{n}{2}}{2} &= n\binom{n}{3} - \binom{n}{4}.\end{aligned}$$

It follows that

$$\begin{aligned}\lambda^2(mn) &= (\lambda^1(m))^2\lambda^2(n) + (\lambda^1(n))^2\lambda^2(m) - 2\lambda^2(m)\lambda^2(n), \\ \lambda^2(\lambda^2(n)) &= \lambda^1(n)\lambda^3(n) - \lambda^4(n).\end{aligned}$$

In general, $\lambda^k(mn)$ is a polynomial with integer coefficients in $\lambda^1(m), \ldots, \lambda^k(m)$, $\lambda^1(n), \ldots, \lambda^k(n)$. Likewise, $\lambda^i(\lambda^j(n))$ is a polynomial with integer coefficients in $\lambda^1(n), \ldots, \lambda^{ij}(n)$.

With the above example in mind, we now give the definition of a λ-ring.

1.2.2 *Basic properties of λ-rings*

Definition 1.10. A *λ-ring* is a ring R together with functions

$$\lambda^n \colon R \to R \quad (n \geq 0),$$

called *λ-operations*, such that for all $x, y \in R$, the following axioms are satisfied:

(1) $\lambda^0(x) = 1$,
(2) $\lambda^1(x) = x$,
(3) $\lambda^n(1) = 0$ for $n \geq 2$,
(4) $\lambda^n(x+y) = \sum_{i+j=n} \lambda^i(x)\lambda^j(y)$,
(5) $\lambda^n(xy) = P_n(\lambda^1(x), \ldots, \lambda^n(x); \lambda^1(y), \ldots, \lambda^n(y))$,
(6) $\lambda^n(\lambda^m(x)) = P_{n,m}(\lambda^1(x), \ldots, \lambda^{nm}(x))$.

Here P_n and $P_{n,m}$ are the universal polynomials with integer coefficients from Example 1.9 and Example 1.7, respectively. If only axioms (1), (2), and (4) are satisfied, then we call R a *pre-λ-ring*.

In a (pre-)λ-ring R, we write

$$\lambda_t(x) = \sum_{n=0}^{\infty} \lambda^n(x)t^n, \tag{1.7}$$

considered as a formal power series in t with coefficients in R. Using this notation, we can rewrite axiom (4) above as

$$\lambda_t(x+y) = \lambda_t(x)\lambda_t(y). \tag{1.8}$$

Definition 1.11. If $\lambda_t(x)$ is a polynomial of degree n, then we say that x has *dimension* n and write $\dim(x) = n$. If every element in R is the difference of finite dimensional elements, then we say that R is *finite dimensional.*

Remark 1.12. What we call a λ-ring (pre-λ-ring) here used to be called a *special λ-ring* (λ-ring) in [Atiyah and Tall (1969); Berthelot *et al.* (1971); Fulton and Lang (1985); Grothendieck (1971)]. Our terminology follows that in [Knutson (1973)]. Also, the dimension of an element in a λ-ring was called the *degree* of that element in [Knutson (1973)].

In Proposition 2.8 and Corollary 2.10 below, we will provide some criteria that make a pre-λ-ring into a λ-ring.

Here are some of the most basic properties of a λ-ring.

Proposition 1.13. *Let R be a λ-ring, and let x and y be elements in R. Then the following statements hold.*

(1) $\lambda_t(1) = 1 + t$.
(2) $\lambda_t(0) = 1$.
(3) $\lambda_t(x+y) = \lambda_t(x)\lambda_t(y)$.

(4) $\lambda_t(-x) = \lambda_t(x)^{-1}$.
(5) $\dim(x+y) \leq \dim(x) + \dim(y)$.
(6) If x and y are both 1-dimensional, then so is xy. So a finite product of 1-dimensional elements is again 1-dimensional.

Proof. All the statements are immediate from the λ-ring axioms. For example, assuming (2) and (3) we have

$$\begin{aligned} 1 &= \lambda_t(0) \\ &= \lambda_t(x - x) \\ &= \lambda_t(x)\lambda_t(-x), \end{aligned}$$

which proves (4).

For statement (6), suppose that $n \geq 2$. In the product

$$h(t) = \prod_{i,j=1}^{n} (1 + x_i y_j t),$$

set $x_2, \ldots, x_n, y_2, \ldots, y_n$ to 0. Then direct inspection shows that the coefficient of t^n is 0. This implies that

$$P_n(s_1, 0, 0, \ldots; \sigma_1, 0, 0, \ldots) = 0,$$

which in turn implies that

$$\lambda^n(xy) = P_n(\lambda^1(x), 0, 0, \ldots; \lambda^1(y), 0, 0, \ldots) = 0.$$

Therefore, the product xy is 1-dimensional. The proofs of the other statements are left as exercises. □

Proposition 1.14. *The following statements hold in a λ-ring R for $k \geq 1$ and $a, a_1, \ldots, a_n \in R$:*

$$\lambda^k(a_1 + \cdots + a_n) = \sum_{i=1}^{n} \lambda^k(a_i) + (\text{polynomial in } \lambda^j(a_l), j < k, 1 \leq l \leq n),$$

$$\lambda^k(-a) = -\lambda^k(a) + (\text{polynomial in } \lambda^j(a), j < k).$$

Here by "polynomial" we mean a polynomial with integer coefficients.

Proof. The first assertion follows from the equality

$$\lambda^k(a_1 + \cdots + a_n) = \sum_{k_1 + \cdots + k_n = k} \lambda^{k_1}(a_1) \cdots \lambda^{k_n}(a_n)$$

and the axiom $\lambda^0(x) = 1$. For the second assertion, we have

$$\begin{aligned} 0 &= \lambda^k(a + (-a)) \\ &= \lambda^k(a) + \lambda^k(-a) + (\text{polynomial in } \lambda^i(a), \lambda^i(-a), i < k). \end{aligned}$$

Thus, we have

$$\lambda^k(-a) = -\lambda^k(a) + (\text{polynomial in } \lambda^i(a), \lambda^i(-a), i < k).$$

Now apply the same procedure to the terms $\lambda^i(-a)$ $(i < k)$, and so forth. The second assertion of the Lemma follows by a finite downward induction. □

1.2.3 *Examples of λ-rings*

Example 1.15. The ring $\mathbf{Z}$ of integers is a finite-dimensional λ-ring with

$$\lambda_t(n) = (1+t)^n$$

for $n \in \mathbf{Z}$, i.e.,

$$\lambda^i(n) = \text{coefficient of } t^i \text{ in } (1+t)^n.$$

Indeed, we have already seen that axioms (1) to (4) are satisfied. Axioms (5) and (6) are certain universal identities about the binomial coefficients. It is possible to prove them directly. However, it is easier to verify them using the universal λ-ring (Corollary 2.11).

Notice that an integer $n \geq 0$ has dimension n, while negative integers do not have finite dimensions.

Example 1.16. The topological K-theory $K(X)$ of any nice-enough space X (e.g., para-compact Hausdorff space) is a λ-ring [Atiyah (1989); Atiyah and Tall (1969)], in which λ^i is induced by the ith exterior power on vector bundles over X. Moreover, the K-theory of a point (or any contractible space) is $K(pt.) \cong \mathbf{Z}$ as a λ-ring. Moreover, if X is a finite CW complex, then $K(X)$ is a finite-dimensional λ-ring.

Example 1.17. For a group G, the complex representation ring $R(G)$ is a λ-ring [Atiyah and Tall (1969)], in which λ^i is induced by the ith exterior power on representations of G. The λ-ring $R(G)$ is finite-dimensional if G is a finite group. Moreover, if $G = \{e\}$ is the trivial group, then $R(\{e\}) \cong \mathbf{Z}$ as a λ-ring. As we will see in §5.5, the trivial group is the *only* finite group whose representation ring is isomorphic to $\mathbf{Z}$. This is a consequence of Knutson's Theorem 5.36.

A lot more examples of λ-rings will be given in the following chapters when we have the universal λ-ring and Adams operations at our disposal. Here is a preview of some of them.

Example 1.18. Let R be a ring. Then its *universal λ-ring* $\Lambda(R)$ is always a λ-ring (Theorem 2.6).

Example 1.19. Let R be a ring. Then the ring $\mathbf{W}(R)$ of *big Witt vectors on* R is a λ-ring, which is isomorphic to $\Lambda(R)$ (Theorem 4.16) via the Artin-Hasse Exponential isomorphism.

Example 1.20. The ring $\hat{\mathbf{Z}}_p$ of p-adic integers is a λ-ring (Corollary 3.63), as is the ring $\mathbf{Z}_{(p)}$ of p-local integers (Corollary 3.59).

Example 1.21. Let G be a group, and let $\mathbf{k}$ be a field of characteristic 0. Let C_G be the set of conjugacy classes of G, and let $\mathbf{k}(C_G)$ be the ring of central functions on G with values in $\mathbf{k}$. Then $\mathbf{k}(C_G)$ is a λ-ring (Chapter 3 Exercise (7d)).

Example 1.22. Every *binomial ring* is a λ-ring (Theorem 5.3). In particular, every ring $\text{Int}(\mathbf{Z}^X)$ of *integer-valued polynomials* is a λ-ring (Theorem 5.21).

Example 1.23. For a binomial ring R, its *necklace ring* $Nr(R)$ is a λ-ring (Corollary 5.45).

Example 1.24. Many quotients of the power series ring $\mathbf{Z}[[x_1, \ldots, x_n]]$ are λ-rings (Theorem 6.13).

1.2.4 *Homomorphisms and ideals*

Definition 1.25. Let R and S be (pre-)λ-rings.

(1) A *(pre-)λ-homomorphism* $f\colon R \to S$ is a ring homomorphism such that $f\lambda^i = \lambda^i f$ for $i \geq 0$.
(2) A *(pre-)λ-ideal* in R is an ideal I in R such that $\lambda^i(x) \in I$ for $i \geq 1$ and $x \in I$.
(3) A *(pre-)λ-subring* of R is a subring R' of R such that $\lambda^i(x) \in R'$ for all $i \geq 0$ and $x \in R'$.

Example 1.26.

(1) If $f\colon X \to Y$ is a map of topological spaces, then the induced map $K(f)\colon K(Y) \to K(X)$ on K-theory is a λ-homomorphism.
(2) If $\varphi\colon G \to H$ is a group homomorphism, then the induced map $R(\varphi)\colon R(H) \to R(G)$ on the representation rings is a λ-homomorphism.

The usual properties and constructions of rings extend in an obvious way to λ-rings.

Proposition 1.27. *Let R and S be λ-rings, and let $f\colon R \to S$ be a λ-homomorphism.*

(1) The kernel of f is a λ-ideal in R.
(2) The image of f is a λ-subring in S.
(3) The quotient R/I of R by a λ-ideal I is naturally a λ-ring, and the projection map $R \to R/I$ is a λ-homomorphism.
(4) The direct product $R \times S$ is a λ-ring, in which

$$\begin{aligned}
\lambda_t(r, 0) &= (1, 1) + \sum_{n \geq 1} (\lambda^n(r), 0)t^n, \\
\lambda_t(0, s) &= (1, 1) + \sum_{n \geq 1} (0, \lambda^n(s))t^n, \\
\lambda_t(r, s) &= \lambda_t(r, 0)\lambda_t(0, s).
\end{aligned}$$

(5) The tensor product $R \otimes S$ is a λ-ring, in which

$$\begin{aligned} \lambda^n(r \otimes 1) &= \lambda^n(r) \otimes 1, \\ \lambda^n(1 \otimes s) &= 1 \otimes \lambda^n(s), \\ \lambda^n(r \otimes s) &= \lambda^n((r \otimes 1)(1 \otimes s)). \end{aligned}$$

(6) If $\{R_\alpha\}$ is an inverse system of λ-rings, then the inverse limit ring $\varprojlim_\alpha R_\alpha$ is naturally a λ-ring.

Proof. Straightforward exercise. □

An ideal in a ring is often described by a set of generators. To check that an ideal I in a λ-ring R is a λ-ideal, one would have to check that $\lambda^n(x) \in I$ for $n \geq 1$ and all $x \in I$. The following result reduces this process to just checking the generators.

Proposition 1.28. *Let R be a λ-ring, and let I be an ideal in R generated by $\{z_j\}_{j \in J}$. Then I is a λ-ideal if and only if $\lambda^n(z_j) \in I$ for $n \geq 1$ and $j \in J$.*

Proof. The "only if" part is obvious. For the other direction, first consider an element in I of the form rz for some $r \in R$ and $z \in \{z_j\}_{j \in J}$. In the polynomial

$$h(t) = \prod_{i,j=1}^{n} (1 + x_i y_j t),$$

if one sets all the y_j to 0, then the coefficient of t^n is 0. This implies that

$$P_n(s_1, \ldots, s_n; 0, 0, \ldots, 0) = 0,$$

which in turn implies that every non-zero term in $P_n(s_1, \ldots, s_n; \sigma_1, \ldots, \sigma_n)$ contains a factor of σ_k for some k. It follows that the element

$$\lambda^n(rz) = P_n(\lambda^1(r), \ldots, \lambda^n(r); \lambda^1(z), \ldots, \lambda^n(z))$$

is a finite sum of terms, each one containing a factor of $\lambda^k(z) \in I$ for some k. So $\lambda^n(rz)$ lies in I as well.

A general element in I is a linear combination $w = \sum_{k=1}^m r_k z_{j_k}$ with coefficients in R. To show that $\lambda^n(w) \in I$ for $n \geq 1$, note that by Proposition 1.13 (3) one has $\lambda_t(w) = \prod_{k=1}^m \lambda_t(r_k z_{j_k})$. Therefore, each $\lambda^n(w)$ is a finite sum of products of $\lambda^p(r_k z_{j_k}) \in I$ for various p and k. This shows that $\lambda^n(w)$ lies in I. □

It was mentioned earlier that $\mathbf{Z}$ is the simplest λ-ring. The following result is the reason for that assertion.

Proposition 1.29. *The following statements hold.*

(1) The ring $\mathbf{Z}$ has a unique λ-ring structure with $\lambda_t(m) = (1+t)^m$ for $m \in \mathbf{Z}$.
(2) Every λ-ring has characteristic 0.
(3) Every λ-ring contains a λ-subring that is isomorphic to $\mathbf{Z}$.

Proof. In any λ-ring, one has $\lambda_t(1) = 1 + t$. Therefore, for a positive integer m one has that

$$\lambda_t(m) = \lambda_t(\underbrace{1 + \cdots + 1}_{m \text{ times}}) = \lambda_t(1)^m = (1+t)^m.$$

Moreover, by Proposition 1.13 (4) one has

$$\lambda_t(-m) = \lambda_t(m)^{-1} = (1+t)^{-m},$$

which shows that the stated λ-ring structure on $\mathbf{Z}$ is the only one.

In a λ-ring R, for a positive integer m, a similar calculation as above shows that

$$\lambda_t(m \cdot 1) = \lambda_t(\underbrace{1 + \cdots + 1}_{m \text{ times}}) = (1+t)^m = 1 + \cdots + t^m.$$

This is a polynomial of degree m over any ring, which shows that $m \neq 0$ in R. Therefore, R has characteristic 0 and has a λ-subring isomorphic to $\mathbf{Z}$. $\square$

1.2.5 *Augmented λ-rings*

Many λ-rings in practice come with a λ-homomorphism to $\mathbf{Z}$. We give these λ-rings a special name.

Definition 1.30. An *augmented λ-ring* is a λ-ring R that comes equipped with a λ-homomorphism $\varepsilon\colon R \to \mathbf{Z}$, called the *augmentation*.

Example 1.31. If x_0 is a base point of a topological space X, then the K-theory $K(X)$ of X is an augmented λ-ring. The augmentation $\varepsilon\colon K(X) \to \mathbf{Z}$ is induced by the map that sends an actual vector bundle over X to the dimension of the fiber over x_0.

Example 1.32. The representation ring $R(G)$ of a group G is an augmented λ-ring. The augmentation $\varepsilon\colon R(G) \to \mathbf{Z}$ is induced by the map that sends an actual representation to its dimension.

Augmented λ-rings admit a natural decomposition with $\mathbf{Z}$ as a factor.

Proposition 1.33. *The following statements hold.*

(1) If R is an augmented λ-ring with augmentation ε, then

$$0 \leq \varepsilon(x) \leq \dim(x)$$

for any finite dimensional element $x \in R$.

(2) A λ-ring R is augmented if and only if there exists a λ-ideal I such that $R = \mathbf{Z} \oplus I$ as an abelian group.

(3) If R and S are augmented λ-rings, then $R \otimes S$ is also an augmented λ-ring.

Proof. If $x \in R$ is of finite dimension d, then for any integer $n > d$ one has that

$$\lambda^n(\varepsilon(x)) = \varepsilon(\lambda^n(x)) = \varepsilon(0) = 0.$$

This shows that $\varepsilon(x) \leq d$ and is non-negative.

For the second statement, suppose that R is augmented with augmentation ε. Let I denote the kernel of ε. Given any element $x \in R$, one can always write it uniquely as

$$x = \varepsilon(x) + (x - \varepsilon(x)),$$

in which $\varepsilon(x) \in \mathbf{Z}$ and $(x - \varepsilon(x)) \in I$. This shows that R decomposes as $\mathbf{Z} \oplus I$. Conversely, if there exists a λ-ideal I such that $R = \mathbf{Z} \oplus I$ as an abelian group, then R is augmented by the projection map onto the first factor.

The tensor product $R \otimes S$ is a λ-ring by Proposition 1.27. If the augmentations of R and S are ε_R and ε_S, respectively, then $R \otimes S$ has augmentation $\varepsilon_R \otimes \varepsilon_S$. □

Given a λ-ring R, it is convenient to know how to put a λ-ring structure on the polynomial ring $R[x]$ or the power series ring $R[[x]]$. This is illustrated by the following result.

Proposition 1.34. *Let R be a λ-ring. Then there exists a unique λ-ring structure on the polynomial ring $R[x]$ such that $\lambda_t(x) = 1 + xt$, i.e.,* $\dim(x) = 1$. *If R is augmented, then so is $R[x]$ with $\varepsilon(x) = 0$ or 1. The same holds for the power series ring $R[[x]]$.*

Proof. We take $\lambda_t(x) = 1 + xt$ as a definition and extend it to all of $R[x]$ or $R[[x]]$ as follows. The λ-operations are extended to the powers x^i using the λ-ring axiom for $\lambda^n(xy)$ repeatedly. For an element $r \in R$, we use this axiom again to extend the λ-operations to monomials of the form rx^i. For a polynomial (or formal power series) $f = \sum_i r_i x^i$ with coefficients in R, we define

$$\lambda_t(f) = \prod_i \lambda_t(r_i x^i)$$

using the axiom for $\lambda^n(x + y)$. The same reasoning shows that the λ-ring structure is uniquely determined by the condition $\lambda_t(x) = 1 + xt$. The assertion about the augmentation is clear. □

We will use this result below in the construction of the free λ-ring on one generator. Applying Proposition 1.34 repeatedly, we obtain the following result.

Corollary 1.35. *Let R be a λ-ring. Then there exists a unique λ-ring structure on the polynomial ring $R[x_1, \ldots, x_n]$ on n variables such that $\lambda_t(x_i) = 1 + x_i t$ for $1 \leq i \leq n$, i.e.,* $\dim(x_i) = 1$ *for each i. If R is augmented, then so is $R[x_1, \ldots, x_n]$ with $\varepsilon(x_i) = 0$ or 1 for each i. The same holds for the power series ring $R[[x_1, \ldots, x_n]]$ on n variables.*

Proof. Apply Proposition 1.34 n times. □

1.3 Free λ-Rings

In this section we construct the free λ-ring U on one generator. This ring U also occurs in (at least) two other places. First, U is also the ring of operations on λ-rings, which we will discuss in §1.4. Second, in representation theory, the ring U is often called the *ring of symmetric functions*. In fact, U is isomorphic to the graded ring

$$R(S) = \bigoplus_{n=0}^{\infty} R(S_n),$$

in which $R(S_n)$ is the representation ring of the symmetric group S_n on n letters [Knutson (1973)].

Before we construct U, let us provide some motivation for its construction. Since U is to be the free λ-ring on one generator, say, x, it should have the following universal property. Given any λ-ring R and an element $r \in R$, there exists a unique λ-homomorphism $u_r \colon U \to R$ such that $u_r(x) = r$. Because U is a λ-ring itself, it should also contain $\lambda^n(x)$ for $n \geq 1$. Therefore, it contains all the polynomials with integer coefficients in x, $\lambda^2(x)$, $\lambda^3(x)$, etc., and hence also the ring

$$U' = \mathbf{Z}[x, \lambda^2(x), \lambda^3(x), \ldots].$$

This latter ring U' is already a λ-ring because of the λ-ring axiom for $\lambda^m(\lambda^n(x))$, which expresses it as a polynomial with integer coefficients in $\lambda^i(x)$ $(1 \leq i \leq mn)$. So U' is a λ-subring of U containing x, which implies that U should be equal to U'.

It turns out that the elements $\lambda^i(x)$ $(i \geq 1)$ in U must also be algebraically independent. In other words, U is a polynomial ring over $\mathbf{Z}$ in countably infinitely many variables $y_1, y_2, \ldots$ such that $\lambda^n(y_1) = y_n$.

We now give the exact construction of U.

1.3.1 *The free λ-ring on one generator*

Let $x_1, x_2, \ldots$ be an infinite sequence of algebraically independent variables. Consider the polynomial ring

$$\Omega_r = \mathbf{Z}[x_1, \ldots, x_r]$$

in $r \geq 0$ variables. Since $\mathbf{Z}$ is a λ-ring (Proposition 1.29), by Corollary 1.35 there exists a unique λ-ring structure on Ω_r in which $\lambda_t(x_i) = 1 + x_i t$, i.e., each x_i has dimension 1.

These λ-rings fit into an inverse system

$$\mathbf{Z} = \Omega_0 \xleftarrow{p_1} \Omega_1 \xleftarrow{p_2} \Omega_2 \xleftarrow{p_3} \Omega_3 \xleftarrow{p_4} \cdots, \tag{1.9}$$

in which the λ-homomorphism

$$p_{r+1} \colon \Omega_{r+1} \to \Omega_r \tag{1.10}$$

is defined by

$$p_{r+1}(x_i) = \begin{cases} x_i & \text{if } 1 \leq i \leq r, \\ 0 & \text{if } i = r+1. \end{cases}$$

Define

$$\Omega = \varprojlim \Omega_r, \tag{1.11}$$

the inverse limit of the inverse system (1.9). Let $\phi_r \colon \Omega \to \Omega_r$ be the structure map that comes with the inverse limit. By Proposition 1.27, Ω is a λ-ring, and ϕ_r is a λ-homomorphism. From the definition of inverse limit, an element in Ω is a power series f with integer coefficients in the variables x_i $(i \geq 1)$ such that for each $r \geq 1$,

$$\phi_r(f) = f(x_1, \ldots, x_r, 0, 0, \ldots)$$

is a polynomial in $x_1, \ldots, x_r$.

Let $s_n(x_1, \ldots, x_r)$ be the nth elementary symmetric function on $x_1, \ldots, x_r$ (Example 1.1). Since

$$\begin{aligned} s_n(x_1, \ldots, x_r, 0) &= p_{r+1}(s_n(x_1, \ldots, x_{r+1})) \\ &= s_n(x_1, \ldots, x_r), \end{aligned}$$

the sequence $\{s_n(x_1, \ldots, x_r)\}_{r \geq 0}$ determines an element

$$s_n = \varprojlim_r s_n(x_1, \ldots, x_r)$$

in Ω. Thinking of s_n as a power series, we have

$$\begin{aligned} s_n &= \sum_{i_1 < \cdots < i_n} x_{i_1} \cdots x_{i_n}, \\ \phi_r(s_n) &= s_n(x_1, \ldots, x_r) = \sum_{1 \leq i_1 < \cdots < i_n \leq r} x_{i_1} \cdots x_{i_n}. \end{aligned} \tag{1.12}$$

For example, s_1 is the infinite sum $s_1 = x_1 + x_2 + x_3 + \cdots$, and $\phi_r(s_1) = x_1 + \cdots + x_r$ for each $r \geq 1$.

In view of the above discussion, we think of s_n as the nth elementary symmetric function on a countable set of variables x_i $(i \geq 1)$. Therefore, we can write down the generating function for these elementary symmetric functions formally as

$$1 + \sum_{n=1}^{\infty} s_n t^n = \prod_{i=1}^{\infty} (1 + x_i t)\,. \tag{1.13}$$

This is the analog of (1.2) involving countably infinitely many variables. There is an obvious analog of The Fundamental Theorem of Symmetric Functions (Theorem 1.4) that involves an infinite sequence of variables x_i $(i \geq 1)$, with s_n (1.12) playing the role of the nth elementary symmetric function.

Proposition 1.36. *In the λ-ring Ω, the elements s_n $(n \geq 1)$ are algebraically independent. Moreover, one has $\lambda^n(s_1) = s_n$ for $n \geq 1$.*

Proof. Suppose that $f(s_1,\ldots,s_n) = 0$ is a polynomial relation involving $s_1,\ldots,s_n$. Then

$$\begin{aligned}0 &= \phi_n f(s_1,\ldots,s_n)\\ &= f(\phi_n(s_1),\ldots,\phi_n(s_n))\\ &= f(s_1(x_1,\ldots,x_n),\ldots,s_n(x_1,\ldots,x_n)).\end{aligned}$$

The algebraic independence of $s_i(x_1,\ldots,x_n)$ $(1 \leq i \leq n)$ (Theorem 1.4) then shows that f is the 0 polynomial. This proves that the elements s_n $(n \geq 1)$ are algebraically independent.

For the second assertion, note that for $r \geq n$, one has

$$\begin{aligned}\phi_r \lambda^n(s_1) &= \lambda^n(\phi_r(s_1))\\ &= \lambda^n(x_1 + \cdots + x_r)\\ &= \sum_{1 \leq i_1 < \cdots < i_n \leq r} x_{i_1} \cdots x_{i_n}\\ &= \phi_r(s_n).\end{aligned}$$

The third equality is a consequence of the λ-ring axiom for $\lambda^n(x + y)$ and the fact that $\dim(x_i) = 1$ for each i. Since this holds for all $r \geq n$, it follows that $\lambda^n(s_1) = s_n$, as desired. □

Definition 1.37. Let U be the smallest λ-subring of Ω that contains s_1. The λ-ring U is called the *free λ-ring on one generator.*

Proposition 1.38. *The free λ-ring on one generator is the polynomial ring*

$$U = \mathbf{Z}[s_1, s_2, s_3, \ldots]$$

on the algebraically independent elements s_i $(i \geq 1)$.

Proof. Since $s_1 \in U$, so is $\lambda^n(s_1) = s_n$ (Proposition 1.36). It follows that U contains the stated polynomial ring. The latter is closed under the λ-operations by the λ-ring axiom on $\lambda^m(\lambda^n(x))$, so it is a λ-ring. Therefore, it must be equal to the ring U. □

We now show that U has the right universal property.

Theorem 1.39. *Let R be a λ-ring, and let r be an element in R. Then there exists a unique λ-homomorphism $u_r \colon U \to R$ such that $\lambda^n(r) = u_r(s_n)$ for $n \geq 1$.*

Proof. Define a ring homomorphism $u_r \colon U \to R$ by first setting $u_r(s_1) = r$. This extends uniquely to a λ-homomorphism on U, since, to be such, u_r must satisfy

$$\lambda^n(r) = \lambda^n(u_r(s_1)) = u_r(\lambda^n(s_1)) = u_r(s_n).$$

So we must choose $u_r(s_n)$ to be $\lambda^n(r)$. □

One can construct the free λ-ring on any set of generators following the procedure above. We leave it to the reader to work out the details in the exercises.

1.4 The Verification Principle

In this section we prove The Verification Principle, which states that a natural operation on λ-rings is uniquely determined by its action on sums of 1-dimensional elements. We begin by defining natural operations on λ-rings.

Definition 1.40. By a *natural operation on λ-rings*, we mean a rule that assigns to each λ-ring R a function (not necessarily a λ-homomorphism) $\mu_R \colon R \to R$ such that, if $f \colon R \to S$ is a λ-homomorphism, then the square

$$\begin{array}{ccc} R & \xrightarrow{f} & S \\ {\scriptstyle \mu_R}\downarrow & & \downarrow{\scriptstyle \mu_S} \\ R & \xrightarrow[f]{} & S \end{array}$$

commutes.

Natural operations on λ-rings can be added and multiplied in the obvious way:

$$\begin{aligned} (\mu + \nu)_R(r) &= \mu_R(r) + \nu_R(r), \\ (\mu\nu)_R(r) &= \mu_R(r)\nu_R(r). \end{aligned}$$

In particular, the set of all natural operations on λ-rings is a ring, which is denoted by Op^λ.

Example 1.41. From the definition of a λ-homomorphism, each λ-operation λ^i $(i \geq 0)$ is a natural operation on λ-rings. It follows that any monomial of the λ-operations, $\lambda^{i_1} \cdots \lambda^{i_n}$, is a natural operation on λ-rings as well. Going one step further, any polynomial with integer coefficients in the λ-operations is a natural operation on λ-rings. We denote this latter ring by $\mathbf{Z}[\lambda^1, \lambda^2, \lambda^3, \ldots]$. Therefore, there is a ring homomorphism

$$\alpha \colon \mathbf{Z}[\lambda^1, \lambda^2, \lambda^3, \ldots] \to Op^\lambda \tag{1.14}$$

given by $\alpha(f(\lambda^1, \ldots, \lambda^n))_R(r) = f(\lambda^1(r), \ldots, \lambda^n(r))$ for an element r in a λ-ring R.

We now show that every natural operation on λ-rings can be expressed uniquely as a polynomial with integer coefficients in the λ-operations.

Theorem 1.42. *The map* α (1.14) *is a ring isomorphism.*

Proof. Since α is clearly a ring homomorphism, it remains to show that α is a bijection. To show that it is injective, suppose that $\alpha(f(\lambda^1, \ldots, \lambda^n)) = 0$. When applied to the free λ-ring on one generator U, one has

$$\begin{aligned} 0 &= \alpha(f(\lambda^1, \ldots, \lambda^n))_U(s_1) \\ &= f(\lambda^1(s_1), \ldots, \lambda^n(s_1)) \\ &= f(s_1, \ldots, s_n). \end{aligned}$$

Since the s_i ($i \geq 1$) are algebraically independent (Proposition 1.36), it follows that $f = 0$, showing that α is injective.

To show that α is surjective, let μ be a natural operation on λ-rings. Applying it to U again, one has $\mu_U(s_1) = g(s_1, \ldots, s_n)$ for some polynomial g with integer coefficients. We claim that

$$\mu = \alpha(g(\lambda^1, \ldots, \lambda^n)). \tag{1.15}$$

Indeed, if r is an element in a λ-ring R, then by Theorem 1.39 there exists a unique λ-homomorphism $u_r \colon U \to R$ such that $u_r(s_n) = \lambda^n(r)$ for all $n \geq 1$. Therefore, we have

$$\begin{aligned} \mu_R(r) &= \mu_R(u_r(s_1)) \\ &= u_r(\mu_U(s_1)) \\ &= u_r(g(s_1, \ldots, s_n)) \\ &= g(u_r(s_1), \ldots, u_r(s_n)) \\ &= g(\lambda^1(r), \ldots, \lambda^n(r)). \end{aligned}$$

This establishes (1.15), showing that α is surjective. □

We are now ready to prove the desired result.

Theorem 1.43 (The Verification Principle). *Every natural operation μ on λ-rings is uniquely a polynomial with integer coefficients in the λ-operations. Moreover, one has*

$$\mu = f(\lambda^1, \ldots, \lambda^n) \tag{1.16}$$

if and only if this equality holds when applied to finite sums of 1-dimensional elements.

Proof. The first assertion was proved in Theorem 1.42. As in the second paragraph of the proof of Theorem 1.42, the condition (1.16) holds if and only if $\mu_U(s_1) = f(s_1, \ldots, s_n)$. It suffices to check this condition in Ω (1.11), which is the inverse limit of Ω_r ($r \geq 0$). Therefore, it suffices to check this condition in each Ω_r:

$$\begin{aligned} \mu_{\Omega_r}(x_1 + \cdots + x_r) &= \mu_{\Omega_r}(\phi_r(s_1)) \\ &= f(\phi_r(s_1), \ldots, \phi_r(s_n)) \\ &= f(s_1(x_1, \ldots, x_r), \ldots, s_n(x_1, \ldots, x_r)). \end{aligned}$$

Since each x_i is 1-dimensional, the Theorem is proved. □

1.5 The Splitting Principle

The purpose of this section is to prove The Splitting Principle, which allows one to write an arbitrary finite dimensional element in a λ-ring as a sum of 1-dimensional elements in a possibly larger λ-ring.

The idea of The Splitting Principle is that, given an n-dimensional element x in a λ-ring R, one would like to write it as a sum

$$x = x_1 + \cdots + x_n$$

of n one-dimensional elements x_i. It proceeds by first adjoining an element x_1 of dimension one to R such that $\dim(x - x_1) = n - 1$. The process is then applied to the $(n-1)$-dimensional element $(x - x_1)$ in $R[x_1]$, and we obtain a 1-dimensional element x_2 such that $\dim(x - x_1 - x_2) = n - 2$. This process is done n times to obtain the desired sum. Moreover, if R is augmented, then each step can be taken to respect the augmentation.

Here is the exact statement of the result.

Theorem 1.44 (The Splitting Principle). *Let x be an n-dimensional element in a λ-ring R. Then there exists a λ-ring S containing R such that*

$$x = x_1 + \cdots + x_n$$

in S, in which each x_i is 1-dimensional. Moreover, if R is augmented with $\varepsilon(x) = m$, then the augmentation can be extended to S in such a way that

$$\varepsilon(x_r) = \begin{cases} 1 & \text{if } 1 \leq r \leq m, \\ 0 & \text{if } m < r \leq n. \end{cases}$$

In order to prove this Theorem, we need some preliminary results.

Lemma 1.45. *The polynomial $P_{n,m}$ (Example 1.7) satisfies*

$$P_{n,m}(s_1, \ldots, s_{m-1}, 0, \ldots, 0) = 0. \tag{1.17}$$

Therefore, every non-zero term in $P_{n,m}(s_1, \ldots, s_{nm})$ contains a factor of s_i for some $i \geq m$.

Proof. Recall that $P_{n,m}$ is the polynomial with integer coefficients in nm variables such that the coefficient of t^n in

$$g(t) = \prod_{1 \leq i_1 < \cdots < i_m \leq nm} (1 + x_{i_1} \cdots x_{i_m} t)$$

is $P_{n,m}(s_1, \ldots, s_{mn})$, where s_i is the ith elementary symmetric function in the x_j's. Setting $x_m = x_{m+1} = \cdots = x_{nm} = 0$ in g, it follows that the coefficient of t^n is 0. This proves (1.17). □

Lemma 1.46. *Let R be a λ-ring, and let x be an n-dimensional element in R. Let y be a variable. In the polynomial λ-ring $R[y]$ in which y has dimension one (Proposition 1.34), the ideal I generated by the element*

$$z = y^n - \lambda^1(x)y^{n-1} + \cdots + (-1)^{n-1}\lambda^{n-1}(x)y + (-1)^n\lambda^n(x)$$

is a λ-ideal.

Proof. By Proposition 1.28, it suffices to show that $\lambda^m(z) \in I$ for $m \geq 1$. To do this, we first claim that I is also generated by $\lambda^n(x-y)$. Indeed, one has

$$\begin{aligned}\lambda_t(x-y) &= \lambda_t(x)\lambda_t(y)^{-1} \qquad \text{(by Proposition 1.13)} \\ &= (1+\lambda^1(x)t+\cdots+\lambda^n(x)t^n)(1+yt)^{-1} \\ &= (1+\lambda^1(x)t+\cdots+\lambda^n(x)t^n)(1-yt+y^2t^2-\cdots).\end{aligned}$$

For $r \geq 0$, the coefficients of t^{n+r} yield the equality

$$\begin{aligned}\lambda^{n+r}(x-y) &= (-1)^{n+r}y^r\left(y^n - \lambda^1(x)y^{n-1} + \cdots + (-1)^n\lambda^n(x)\right) \\ &= (-1)^{n+r}y^r z.\end{aligned} \tag{1.18}$$

Setting $r=0$ we see that $z = (-1)^n\lambda^n(x-y)$, so I is also generated by $\lambda^n(x-y)$.

To prove the Lemma, it suffices to show that $\lambda^m(\lambda^n(x-y)) \in I$ for $m \geq 1$. Now

$$\lambda^m(\lambda^n(x-y)) = P_{m,n}(\lambda^1(x-y),\ldots,\lambda^{mn}(x-y))$$

is a sum of terms, each one containing a factor of $\lambda^{n+r}(x-y)$ for some $r \geq 0$ (by Lemma 1.45). Since $\lambda^{n+r}(x-y) \in I$ by (1.18), we conclude that $\lambda^m(\lambda^n(x-y)) \in I$ as well. □

Lemma 1.47. *In the setting of Theorem 1.44, there exists a λ-ring $R[x_1]$ containing R such that* $\dim(x_1) = 1$ *and* $\dim(x - x_1) = n-1$. *Moreover, if R is augmented with $\varepsilon(x) = m$, then the augmentation can be extended to $R[x_1]$ in such a way that*

$$\varepsilon(x_1) = \begin{cases} 1 & \text{if } m > 0, \\ 0 & \text{if } m = 0. \end{cases}$$

In particular, one has

$$\varepsilon(x - x_1) = \begin{cases} m-1 & \text{if } m > 0, \\ 0 & \text{if } m = 0. \end{cases}$$

Proof. Consider the quotient λ-ring $R[x_1] = R[y]/I$, in which y and I are as in Lemma 1.46 and x_1 is the image of y in the quotient. The element x_1 has dimension 1 because y does. Moreover, the element $(x - x_1)$ has dimension $(n-1)$ because $\lambda^{n+r}(x-y) \in I$ for $r \geq 0$ by (1.18) and $\lambda^{n-1}(x-y) \notin I$

Now suppose that R is augmented. Recall that the λ-ring $R[y]$ can be augmented by either $\varepsilon(y) = 0$ or $\varepsilon(y) = 1$ by Proposition 1.34. If $\varepsilon(I) = 0$, then $R[x_1]$ is augmented by $\varepsilon(w + I) = \varepsilon(w)$ for $w \in R[y]$. If $\varepsilon(x) = m$, then

$$\varepsilon(\lambda^r(x)) = \lambda^r(\varepsilon(x)) = \binom{m}{r}.$$

Now if $m > 0$, we augment $R[y]$ by setting $\varepsilon(y) = 1$. This implies that

$$\begin{aligned}\varepsilon(z) &= \varepsilon(y^n - \lambda^1(x)y^{n-1} + \cdots + (-1)^{n-1}\lambda^{n-1}(x)y + (-1)^n\lambda^n(x)) \\ &= 1 - \binom{m}{1} + \binom{m}{2} + \cdots + (-1)^m\binom{m}{m} \\ &= (1-1)^m \\ &= 0.\end{aligned}$$

The second equality uses the fact that $\varepsilon(x) = m \leq n = \dim(x)$ (by Proposition 1.33). Therefore, $\varepsilon(I) = 0$, and we can augment $R[x_1]$ by setting $\varepsilon(x_1) = \varepsilon(y) = 1$.

On the other hand, if $m = 0$, then we augment $R[y]$ by setting $\varepsilon(y) = 0$. This implies that

$$\begin{aligned}\varepsilon(z) &= \varepsilon(y^n - \lambda^1(x)y^{n-1} + \cdots + (-1)^{n-1}\lambda^{n-1}(x)y + (-1)^n\lambda^n(x)) \\ &= 0.\end{aligned}$$

Since $\varepsilon(I) = 0$, we obtain the induced augmentation $\varepsilon(x_1) = \varepsilon(y) = 0$ on $R[x_1]$. □

Proof. [Proof of Theorem 1.44] Perform a downward induction on $n = \dim(x)$ using Lemma 1.47. We obtain a λ-ring $S = R[x_1, \ldots, x_n]$ in which each x_i has dimension 1 and $(x - x_1 - \cdots - x_n)$ has dimension 0, i.e., $x = x_1 + \cdots + x_n$, as desired. □

Corollary 1.48. *Let x be an n-dimensional element in a λ-ring R. Then $\lambda^n(x)$ is 1-dimensional.*

Proof. Extending R to S if necessary as in Theorem 1.44, we can write x as the sum of n one-dimensional elements x_i $(1 \leq i \leq n)$. Then

$$\begin{aligned}\lambda^n(x) &= \lambda^n(x_1 + \cdots + x_n) \\ &= x_1 \cdots x_n,\end{aligned}$$

which is 1-dimensional because each x_i is (Proposition 1.13 (6)). □

1.6 Exercises

(1) Let $f(t) = 1 + t + a_2t^2 + a_3t^3 + \cdots$ be a power series with integer coefficients. Show that $\lambda_t(n) = f(t)^n$ defines a pre-λ-ring structure on $\mathbf{Z}$.
(2) Prove Theorem 1.8.
(3) Write the coefficient of t^k $(2 \leq k \leq 5)$ in

$$g(t) = \prod_{1 \leq i < j \leq 4} (1 + x_i x_j t)$$

as a polynomial in the elementary symmetric functions on x_1, x_2, x_3, x_4. See Example 1.7.
(4) Write down the polynomials $P_{2,3}$ and $P_{3,2}$ (Example 1.7) explicitly.
(5) Write down the polynomials P_3 and P_4 (Example 1.9) explicitly.
(6) Prove that $P_n(a_1, \ldots, a_n; b_1, \ldots, b_n)$ is a polynomial in which every non-zero monomial contains a factor of a_i and a factor of b_j for some $i, j \in \{1, \ldots, n\}$.
(7) Let $s_1, \ldots, s_n$ be the elementary symmetric functions of the variables $x_1, \ldots, x_n$. Prove:
  (a) Each x_i^n lies in the ideal $\langle s_1, \ldots, s_n \rangle$ generated by $s_1, \ldots, s_n$ in the polynomial ring $\mathbf{Z}[x_1, \ldots, x_n]$.

(b) Let $f \in \mathbf{Z}[x_1, \ldots, x_n]$ be a homogeneous polynomial of degree $> n(n-1)$. Then f lies in the ideal $\langle s_1, \ldots, s_n \rangle$.

(8) Finish the proof of Proposition 1.13.

(9) Prove Proposition 1.27.

(10) Check that the map $p_{r+1}\colon \Omega_{r+1} \to \Omega_r$ in (1.10) is actually a λ-homomorphism.

(11) Modify the constructions in §1.3 to construct the free λ-ring U_2 on two generators s_1 and σ_1. Show that it has the right universal property, analogous to Theorem 1.39.

(12) Define a natural operation μ with two inputs on λ-rings as a rule that assigns to every λ-ring R a function $\mu_R\colon R \times R \to R$ that is natural with respect to λ-homomorphisms. Write down the exact definition. Then prove the analogue of The Verification Principle (Theorem 1.43) for such operations.

Chapter 2

Universal λ-Rings

In this Chapter we discuss the universal λ-ring $\Lambda(R)$ of a ring R. This construction provides a more conceptual definition of a λ-ring. Indeed, a λ-ring structure on a ring R is equivalent to a ring homomorphism $R \to \Lambda(R)$ that satisfies some additional properties. The λ-ring $\Lambda(R)$ also makes it easier to check whether a ring R is a λ-ring. The assignment that sends a ring R to the λ-ring $\Lambda(R)$ is a functor. In fact, Λ is a *comonad*, which is a functor together with a comultiplication of some sort.

In §2.1 we construct the universal λ-ring $\Lambda(R)$ of R (Theorem 2.6) and show that it has certain nice properties. To prepare for the discussion of Λ as a comonad, in §2.2 we introduce the language of categories and functors, and in §2.3 we discuss comonads and their coalgebras. In §2.4 we consider the comonad Λ associated to the universal λ-ring construction (Theorem 2.25) and describe λ-rings as Λ-coalgebras (Theorem 2.26).

There is another important description of the universal λ-ring. Namely, $\Lambda(R)$ is isomorphic to the ring $\mathbf{W}(R)$ of big Witt vectors on R. We will discuss Witt vectors in Chapter 4 when we have Adams operations (Chapter 3) at our disposal.

2.1 The Universal λ-Ring $\Lambda(R)$

In this section, we discuss the universal λ-ring $\Lambda(R)$ of a ring R. The original references for the universal λ-ring are [Berthelot (1971); Grothendieck (1971)]. Other references for the universal λ-rings are [Atiyah and Tall (1969); Hazewinkel (1978); Knutson (1973); Patras (2003)].

We first discuss the ring structure on $\Lambda(R)$ (Theorem 2.5). Then we equip it with λ-operations and show that it is a λ-ring (Theorem 2.6).

2.1.1 $\Lambda(R)$ *is a ring*

Let R be a ring.

Definition 2.1. By a *strict power series* over R, we mean a formal power series with coefficients in R of the form $f(t) = 1 + a_1 t + a_2 t^2 + a_3 t^3 + \cdots$, i.e., it has constant term 1.

Remark 2.2. Such power series are sometimes called *special* in the literature (e.g., [Knutson (1973)]).

The following Proposition shows that every strict power series has a multiplicative inverse.

Proposition 2.3. *Let $f(t) = 1 + \sum_{n \geq 1} a_n t^n$ be a strict power series over R. Then there exists a unique strict power series $f(t)^{-1} = 1 + \sum_{n \geq 1} b_n t^n$ such that $f(t)f(t)^{-1} = 1$. Moreover, the coefficient b_m is a polynomial with integer coefficients in $a_1, \ldots, a_m$ of the form*

$$b_m = -a_m + (\textit{terms involving } a_1, \ldots, a_{m-1}). \tag{2.1}$$

Proof. The multiplicative inverse $f(t)^{-1}$ of $f(t)$ can be constructed by induction. Take $b_1 = -a_1$. Then we have

$$f(t)(1 + b_1 t) \equiv 1 \pmod{t^2}.$$

Inductively, suppose that $b_1, \ldots, b_{n-1}$ have been found such that (2.1) is satisfied for $1 \leq m < n$ and that

$$f(t)(1 + b_1 t + \cdots + b_{n-1} t^{n-1}) \equiv 1 \pmod{t^n}.$$

Then

$$b_n = -a_1 b_{n-1} - a_2 b_{n-2} - \cdots - a_{n-1} b_1 - a_n$$

clearly satisfies (2.1). Moreover, the coefficient of t^n in

$$f(t)(1 + b_1 t + \cdots + b_{n-1} t^{n-1} + b_n t^n)$$

is

$$b_n + a_1 b_{n-1} + a_2 b_{n-2} + \cdots + a_{n-1} b_1 + a_n = 0.$$

It follows that

$$f(t)(1 + b_1 t + \cdots + b_{n-1} t^{n-1} + b_n t^n) \equiv 1 \pmod{t^{n+1}}.$$

Therefore, by induction $f(t)^{-1}$ exists. Since each coefficient b_n is uniquely determined by a_i $(1 \leq i \leq n)$, the uniqueness of $f(t)^{-1}$ follows. □

Proposition 2.3 can be rephrased as follows. There exist universal polynomials with integer coefficients i_n $(n \geq 1)$ in n variables such that

$$i_n(x_1, \ldots, x_n) = -x_n + (\text{terms involving } x_1, \ldots, x_{n-1}) \tag{2.2}$$

and

$$\left(1 + a_1 t + a_2 t^2 + a_3 t^3 + \cdots\right)^{-1} = 1 + \sum_{m \geq 1} i_m(a_1, \ldots, a_m) t^m.$$

For example, we have

$$\begin{aligned} i_1(x_1) &= -x_1, \\ i_2(x_1,x_2) &= -x_2 + x_1^2, \\ i_3(x_1,x_2,x_3) &= -x_3 + 2x_1x_2 - x_1^3. \end{aligned} \tag{2.3}$$

Moreover, if we assign the degree n to x_n, then i_n is a homogeneous polynomial of degree n.

By Proposition 2.3 the set of all strict power series over R is an abelian group in which addition is given by multiplication of power series. The additive 0 is the strict power series 1. We first extend this structure to a ring.

Definition 2.4. Let $\Lambda(R)$ denote the set of all strict power series over R.

Define the following operations on $\Lambda(R)$. For strict power series

$$f = f(t) = 1 + \sum_{n\geq 1} a_n t^n \quad \text{and} \quad g = g(t) = 1 + \sum_{n\geq 1} b_n t^n$$

over R, set:

$$\begin{aligned} f +_\Lambda g &= \Big(1 + \sum_{n\geq 1} a_n t^n\Big)\Big(1 + \sum_{n\geq 1} b_n t^n\Big), \\ f \times_\Lambda g &= 1 + \sum_{n\geq 1} P_n(a_1,\ldots,a_n; b_1,\ldots,b_n)t^n. \end{aligned} \tag{2.4}$$

The polynomial P_n is the same one that appeared in the definition of a λ-ring and was defined in Example 1.9.

Note that $+_\Lambda$ is simply multiplication of strict power series. The multiplication $\times_\Lambda$ is clearly commutative and has an identity. Namely, it is the strict power series $1+t$. Indeed, in the polynomial

$$h(t) = \prod_{i,j=1}^{n} (1 + x_i y_j t),$$

if we set $y_2 = \cdots = y_n = 0$, then the coefficient of t^n in $h(t)$ becomes

$$(x_1 \cdots x_n)y_1^n = s_n \sigma_1^n.$$

It follows that

$$P_n(a_1,\ldots a_n; b_1, 0, 0, \ldots 0) = a_n b_1^n.$$

So we have

$$\Big(1 + \sum_{n\geq 1} a_n t^n\Big) \times_\Lambda (1 + b_1 t) = 1 + \sum_{n\geq 1} a_n b_1^n t^n, \tag{2.5}$$

which implies that $1+t$ is the identity for the operation $\times_\Lambda$.

Theorem 2.5. *Let R be a ring. Then the set $\Lambda(R)$ is a ring with*

- *addition* $+_\Lambda$,
- *multiplication* $\times_\Lambda$,
- *additive zero* 1, *and*
- *multiplicative identity* $1+t$.

Proof. It follows from Proposition 2.3 that $\Lambda(R)$ is an abelian group under $+_\Lambda$ with additive zero 1. From the discussion just before the Theorem, to show that $\Lambda(R)$ is a ring, it remains to check the following axioms for $e, f, g \in \Lambda(R)$:

$$(e \times_\Lambda f) \times_\Lambda g = e \times_\Lambda (f \times_\Lambda g), \tag{2.6a}$$

$$e \times_\Lambda (f +_\Lambda g) = (e \times_\Lambda f) +_\Lambda (e \times_\Lambda g). \tag{2.6b}$$

For (2.6a) it suffices to prove it in the universal case, i.e., when the coefficients in e, f, and g are algebraically independent. In particular, we may assume that the coefficients in e, f, and g are the elementary symmetric functions (which are algebraically independent) in three countably infinite sets of variables x_i, y_i, and z_i $(i \geq 0)$, respectively. So we write formally

$$e = \prod_{i\geq 1}(1+x_i t), \quad f = \prod_{j\geq 1}(1+y_j t), \quad \text{and} \quad g = \prod_{k\geq 1}(1+z_k t), \tag{2.7}$$

as in (1.13). The associativity condition (2.6a) now follows from the equalities:

$$\begin{aligned}
(e \times_\Lambda f) \times_\Lambda g &= \prod_{i,j}(1+x_i y_j t) \times_\Lambda \prod_k (1+z_k t) \\
&= \prod_{i,j,k}(1+x_i y_j z_k t) \\
&= \prod_i (1+x_i t) \times_\Lambda \prod_{j,k}(1+y_j z_k t) \\
&= e \times_\Lambda (f \times_\Lambda g).
\end{aligned}$$

For the distributive property (2.6b), we use the same setting (2.7) as in the previous paragraph. Then we have

$$f +_\Lambda g = \prod_{j,k}(1+y_j t)(1+z_k t).$$

The left-hand side of (2.6b) is

$$\begin{aligned}
e \times_\Lambda (f +_\Lambda g) &= \prod_i (1+x_i t) \times_\Lambda \prod_{j,k}(1+y_j t)(1+z_k t) \\
&= \prod_{i,j}(1+x_i y_j t) \cdot \prod_{i,k}(1+x_i z_k t) \\
&= (e \times_\Lambda f) +_\Lambda (e \times_\Lambda g).
\end{aligned}$$

This proves (2.6b). □

2.1.2 $\Lambda(R)$ *is a* λ*-ring*

Now we put a λ-ring structure on the ring $\Lambda(R)$. For $m \geq 1$ define

$$\lambda^m(f) = 1 + \sum_{n \geq 1} P_{n,m}(a_1, \ldots, a_{nm})t^n \tag{2.8}$$

for a strict power series $f = 1 + \sum_{n\geq 1} a_n t^n \in \Lambda(R)$. Here $P_{n,m}$ is the same polynomial that appeared in the definition of a λ-ring and was defined in Example 1.7. Also, we set

$$\lambda^0(f) = 1 + t \tag{2.9}$$

for any $f \in \Lambda(R)$.

Theorem 2.6. *Let R be a ring. When equipped with the operations λ^m $(m \geq 0)$, the ring $\Lambda(R)$ becomes a λ-ring.*

Proof. We need to check axioms (2) to (6) in Definition 1.10, since axiom (1) holds by definition (2.9). To check that λ^1 is the identity map (axiom (2)), note that

$$\lambda^1(f) = 1 + \sum_{n\geq 1} P_{n,1}(a_1, \ldots, a_n)t^n = 1 + \sum_{n \geq 1} a_n t^n$$

by (1.5).

For axiom (3), observe that for $m \geq 2$, we have

$$\lambda^m(1 + at) = 1 + \sum_{n\geq 1} P_{n,m}(a, 0, 0, \ldots, 0)t^n.$$

By Lemma 1.45 we know that every non-zero term in $P_{n,m}(a_1, \ldots, a_{nm})$ must contain a factor of a_i for some $i \geq m$. Since $m \geq 2$ we have $P_{n,m}(a, 0, \ldots, 0) = 0$, so, in particular, $\lambda^m(1 + at) = 1$. The case $a = 1$ shows that axiom (3) is true.

For axiom (4) in Definition 1.10, we use the notation $\sum_\Lambda$ to denote an indexed sum in $\Lambda(R)$. As in the proof of Theorem 2.5, it suffices to prove

$$\lambda^n (f +_\Lambda g) = \sum_{l=0}^{n}{}_{\Lambda} \left(\lambda^l(f) \times_\Lambda \lambda^{n-l}(g)\right) \tag{2.10}$$

when the coefficients of f and g are the elementary symmetric functions in the variables x_i and y_j $(i, j \geq 1)$, respectively. So we write formally

$$f = \prod_{i\geq 1}(1 + x_i t) \quad \text{and} \quad g = \prod_{j \geq 1}(1 + y_j t). \tag{2.11}$$

Then we have

$$\lambda^l(f) = \prod_{i_1 < \cdots < i_l} (1 + x_{i_1} \cdots x_{i_l} t).$$

Since

$$f +_\Lambda g = \prod_{i,j}(1 + x_i t)(1 + y_j t),$$

we have

$$\lambda^n(f +_\Lambda g) = \prod_{\substack{0\le l\le n \\ i_1<\cdots<i_l \\ j_1<\cdots<j_{n-l}}} \left(1 + x_{i_1}\cdots x_{i_l} y_{j_1}\cdots y_{j_{n-l}} t\right). \tag{2.12}$$

On the other hand, we have

$$\begin{aligned} &\lambda^l(f) \times_\Lambda \lambda^{n-l}(g) \\ &= \prod_{i_1<\cdots<i_l} (1 + x_{i_1}\cdots x_{i_l} t) \times_\Lambda \prod_{j_1<\cdots<j_{n-l}} (1 + y_{j_1}\cdots y_{j_{n-l}} t) \\ &= \prod_{\substack{i_1<\cdots<i_l \\ j_1<\cdots<j_{n-l}}} \left(1 + x_{i_1}\cdots x_{i_l} y_{j_1}\cdots y_{j_{n-l}} t\right). \end{aligned}$$

Summing the last expression in $\Lambda(R)$ for $0 \le l \le n$ gives the expression for $\lambda^n(f +_\Lambda g)$ in (2.12). This proves (2.10).

Note that at this point, we have shown that $\Lambda(R)$ is a pre-λ-ring. In particular, by axiom (4) which was just proved in the previous paragraph, the map $\lambda_t \colon \Lambda(R) \to \Lambda(\Lambda(R))$ is additive. It is easy to see that the 1-dimensional elements in $\Lambda(R)$ are exactly the linear polynomials. Moreover, the product of two 1-dimensional elements in $\Lambda(R)$ is again 1-dimensional, since

$$(1 + at) \times_\Lambda (1 + bt) = 1 + abt, \tag{2.13}$$

To prove axiom (5) in Definition 1.10, first observe that it is equivalent to the condition

$$\lambda_t(f \times_\Lambda g) = \lambda_t(f) \times_\Lambda \lambda_t(g) \tag{2.14}$$

in $\Lambda(\Lambda(R))$ for f and g in $\Lambda(R)$. It follows from (2.13) that (2.14) holds when both f and g are 1-dimensional elements. For the general case, we write f and g as in (2.11),

$$\begin{aligned} f &= \prod_{i\ge1}(1 + x_i t) = \sum\nolimits_{\Lambda}{}_{i\ge1} \alpha_i, \\ g &= \prod_{j\ge1}(1 + y_j t) = \sum\nolimits_{\Lambda}{}_{j\ge1} \beta_j, \end{aligned} \tag{2.15}$$

where

$$\alpha_i = 1 + x_i t \quad\text{and}\quad \beta_j = 1 + y_j t$$

are 1-dimensional elements. Using the additivity of λ_t, we have

$$\begin{aligned}
\lambda_t(f \times_\Lambda g) &= \lambda_t\Big({\sum_{i,j}}_\Lambda (\alpha_i \times_\Lambda \beta_j)\Big) \\
&= {\sum_{i,j}}_\Lambda \lambda_t(\alpha_i \times_\Lambda \beta_j) \\
&= {\sum_{i,j}}_\Lambda (\lambda_t(\alpha_i) \times_\Lambda \lambda_t(\beta_j)) \\
&= \Big({\sum_i}_\Lambda \lambda_t(\alpha_i)\Big) \times_\Lambda \Big({\sum_j}_\Lambda \lambda_t(\beta_j)\Big) \\
&= \lambda_t\Big({\sum_i}_\Lambda \alpha_i\Big) \times_\Lambda \lambda_t\Big({\sum_j}_\Lambda \beta_j\Big) \\
&= \lambda_t(f) \times_\Lambda \lambda_t(g).
\end{aligned}$$

This proves (2.14).

The proof of axiom (6) is similar to the proof of axioms (4) and (5) above. Indeed, axiom (6) in $\Lambda(R)$ is equivalent to the condition

$$\lambda_t(\lambda^m(f)) = \lambda^m(\lambda_t(f)) \in \Lambda(\Lambda(R)) \tag{2.16}$$

for $f \in R$ and $m \geq 1$. As above, first observe that this is true if f has dimension 1. Then extend it to the general case in which f is a sum (in $\Lambda(R)$) of 1-dimensional element as in (2.15). We leave the details to the reader as an exercise. □

Definition 2.7. The λ-ring $\Lambda(R)$ is called the *universal λ-ring* of R.

The universal λ-ring has several nice properties. Recall from (1.7) that for a pre-λ-ring R, we defined a map $\lambda_t\colon R \to \Lambda(R)$ by

$$\lambda_t(a) = 1 + \sum_{n\geq 1} \lambda^n(a)t^n$$

for $a \in R$.

Proposition 2.8. *Let R be a pre-λ-ring (Definition 1.10). Then R is a λ-ring if and only if the map $\lambda_t\colon R \to \Lambda(R)$ is a pre-λ-homomorphism (Definition 1.25).*

Proof. Recall from (1.8) that the condition

$$\lambda_t(a+b) = \lambda_t(a)\lambda_t(b) = \lambda_t(a) +_\Lambda \lambda_t(b)$$

holds in any pre-λ-ring. Thus, the map λ_t is additive. Moreover, it clearly sends 0 in R to 1, the additive 0 in $\Lambda(R)$. The multiplicativity of λ_t is equivalent to axiom (5) in Definition 1.10. The map λ_t preserves the multiplicative identity if and only if $\lambda_t(1) = 1 + t$, which is equivalent to axiom (3) in Definition 1.10. Finally, the compatibility of λ_t with λ^m is equivalent to axiom (6) in Definition 1.10. □

Proposition 2.9. *In the universal λ-ring $\Lambda(R)$ of a ring R, the following statements hold.*

(1) An element $f \in \Lambda(R)$ has dimension n if and only if it is a polynomial of degree n. In particular, 1-dimensional elements in $\Lambda(R)$ are the linear polynomials $1 + at$ with $a \neq 0$.
(2) If $f = 1 + a_1 t + \cdots + a_n t^n \in \Lambda(R)$ with $a_n \neq 0$, then $\lambda^n(f) = 1 + a_n t$.
(3) One has

$$\left(1 + \sum_{n \geq 1} a_n t^n\right) \times_\Lambda (1 + bt) = 1 + \sum_{n \geq 1} a_n b^n t^n.$$

Proof. As in (2.15) we write an element f as $f = \prod_i (1 + x_i t)$. The first two assertions then follow from the identity

$$\begin{aligned}\lambda_t\Big(\prod_i (1 + x_i t)\Big) &= \lambda_t\Big(\sum_{i}{}_{\Lambda}(1 + x_i t)\Big) \\ &= \prod_i \lambda_t(1 + x_i t)\end{aligned}$$

and the fact that $(1 + x_i t)$ has dimension 1. The last assertion was already proved in (2.5). □

Corollary 2.10. *Let R be a pre-λ-ring in which the following three conditions are satisfied:*

(1) $\lambda_t(1) = 1 + t$.
(2) Every element $r \in R$ can be written as a finite sum $r = \sum \pm a_i$ in which each a_i has dimension 1.
(3) The product of two 1-dimensional elements is again 1-dimensional.

Then R is a λ-ring.

Proof. In view of Proposition 2.8, it remains to show that the map λ_t is multiplicative and is compatible with the λ^n.

Since the map λ_t is additive, using the hypotheses on R, to check that

$$\lambda_t(ab) = \lambda_t(a) \times_\Lambda \lambda_t(b)$$

for a and b in R, we may assume that a and b have dimension 1. In this case, the product ab also has dimension 1, so we have

$$\begin{aligned}\lambda_t(ab) &= 1 + abt \\ &= (1 + at) \times_\Lambda (1 + bt) \\ &= \lambda_t(a) \times_\Lambda \lambda_t(b).\end{aligned}$$

This shows that λ_t is multiplicative.

Finally, to show that

$$\lambda^n(\lambda_t(a)) = \lambda_t(\lambda^n(a)) \tag{2.17}$$

for $a \in R$, first observe that by Proposition 2.9, this is true if a has dimension 1. For the general case, using the hypothesis on R, it suffices to show that if (2.17) holds for a and b, then it holds for $a + b$. We have

$$\begin{aligned}
\lambda_t(\lambda^n(a+b)) &= \lambda_t\Big(\sum_{i=0}^{n} \lambda^i(a)\lambda^{n-i}(b)\Big) \\
&= \sum_{i=0}^{n}{}_{\Lambda}\, \lambda_t\left(\lambda^i(a)\lambda^{n-i}(b)\right) \\
&= \sum_{i=0}^{n}{}_{\Lambda}\, \lambda_t(\lambda^i(a)) \times_{\Lambda} \lambda_t(\lambda^{n-i}(b)) \\
&= \sum_{i=0}^{n}{}_{\Lambda}\, \lambda^i(\lambda_t(a)) \times_{\Lambda} \lambda^{n-i}(\lambda_t(b)) \\
&= \lambda^n\left(\lambda_t(a) +_{\Lambda} \lambda_t(b)\right) \\
&= \lambda^n(\lambda_t(a+b)).
\end{aligned}$$

This shows that (2.17) holds for $a + b$. □

Corollary 2.11. *The ring* $\mathbf{Z}$ *of integers is a* λ*-ring with* $\lambda_t(n) = (1+t)^n$ *for* $n \in \mathbf{Z}$.

Proof. Apply Corollary 2.10 to $\mathbf{Z}$. Indeed, it is clear that

$$\lambda_t(m+n) = (1+t)^{m+n} = \lambda_t(m)\lambda_t(n).$$

Thus, $\mathbf{Z}$ is a pre-λ-ring. The three conditions in Corollary 2.10 are clearly satisfied for $\mathbf{Z}$, which implies that $\mathbf{Z}$ is a λ-ring. □

The construction of the universal λ-ring $\Lambda(R)$ of a ring R has other good formal properties. In preparation for the discussion in §2.4 and Chapter 4 of these properties, we first introduce the formal language of category theory.

2.2 Categories and Functors

Mathematical objects often come with the basic operation of composition of maps. For example, take the collection of groups. Every group G has an identity homomorphism 1_G, and given any two groups G_1 and G_2, there is a set of group homomorphisms from G_1 to G_2. Given group homomorphisms $f\colon G_1 \to G_2$ and $g\colon G_2 \to G_3$, there is a composition $gf\colon G_1 \to G_3$. Composition of homomorphisms satisfies two properties:

(1) $f1_G = f$ and $1_G g = g$ for $f\colon G \to H$ and $g\colon H \to G$.
(2) $(hg)f = h(gf)$ for $f\colon G_1 \to G_2$, $g\colon G_2 \to G_3$, and $h\colon G_3 \to G_4$.

The same can be said of other objects, such as sets, abelian groups, rings, modules over a ring, and so forth.

This notion of composition of maps is formalized in the language of category theory. In this section, we will give some basic definitions in category theory. Our aim here is to provide the framework for the following sections when we discuss comonads and their coalgebras. A general reference for category theory is [Mac Lane (1998)]. The book [Lang (2002)] also has most of the basic definitions that we need.

2.2.1 *Categories*

A *category* $\mathcal{C}$ consists of:

(1) a class $Ob(\mathcal{C})$ of *objects*,
(2) a set $\mathcal{C}(A, B)$ of *morphisms* for every pair (A, B) of objects in $\mathcal{C}$,
(3) an *identity morphism* $1_A \in \mathcal{C}(A, A)$ for every object A in $\mathcal{C}$, and
(4) a *composition function*

$$\mathcal{C}(A, B) \times \mathcal{C}(B, C) \to \mathcal{C}(A, C)$$

for every triple (A, B, C) of objects in $\mathcal{C}$.

If f is a morphism in $\mathcal{C}(A, B)$, we write it as

$$f \colon A \to B.$$

For morphisms $f \colon A \to B$ and $g \colon B \to C$, we write $gf \colon A \to C$ for their composition. There are two axioms that need to be satisfied:

Associativity Composition of morphisms is associative, i.e., $(hg)f = h(gf)$ for morphisms $f \colon A \to B$, $g \colon B \to C$, and $h \colon C \to D$.
Unit Given any morphism $f \colon A \to B$, $1_B f = f = f 1_A$.

A morphism $f \colon A \to B$ is said to be an *isomorphism* if there exists a morphism $g \colon B \to A$ such that $gf = 1_A$ and $fg = 1_B$.

Below is a list of familiar examples of categories:

(1) **Set**: The category whose objects are sets and whose morphisms are functions.
(2) **Top**: The category whose objects are topological spaces and whose morphisms are continuous functions.
(3) **Group**: The category whose objects are groups and whose morphisms are group homomorphisms.
(4) **Ab**: The category whose objects are abelian groups and whose morphisms are group homomorphisms.
(5) **Ring**: The category whose objects are commutative rings with identity and whose morphisms are ring homomorphisms.
(6) $\mathbf{Ring}^{\lambda}$: The category whose objects are λ-rings and whose morphisms are λ-homomorphisms.

2.2.2 *Functors*

Functors give us a way to compare and relate categories. Let $\mathcal{C}$ and $\mathcal{D}$ be categories. A *functor* $F: \mathcal{C} \to \mathcal{D}$ is an assignment that assigns to every object A in $\mathcal{C}$ an object $F(A)$ in $\mathcal{D}$ and to every morphism $f: A \to B$ in $\mathcal{C}$ a morphism $F(f): F(A) \to F(B)$ in $\mathcal{D}$. It is required to preserve the identity morphisms and compositions. In other words:

Identity For every object A in $\mathcal{C}$, $F(1_A) = 1_{F(A)}$.
Composition If $f: A \to B$ and $g: B \to C$ are morphisms in $\mathcal{C}$, then $F(gf) = F(g)F(f)$ in $\mathcal{D}$.

Given two functors $F: \mathcal{C} \to \mathcal{D}$ and $G: \mathcal{D} \to \mathcal{E}$, their *composition* $GF: \mathcal{C} \to \mathcal{E}$ is the functor defined by

$$\begin{aligned}(GF)(A) &= G(F(A)), \\ (GF)(f) &= G(F(f))\end{aligned}$$

for an object A and a morphism f in $\mathcal{C}$.

If $F: \mathcal{C} \to \mathcal{C}$ is a functor whose domain and target categories are the same and if $n \geq 1$, then we define the functor

$$F^n = \underbrace{F \cdots F}_{n \text{ times}}: \mathcal{C} \to \mathcal{C},$$

the composition of F with itself n times.

Example 2.12. There is a functor U: $\mathbf{Ring}^\lambda \to \mathbf{Ring}$ that assigns to every λ-ring its underlying ring and every λ-homomorphism its underlying ring homomorphism. In other words, U simply forgets about the λ-operations. Such a functor is called a *forgetful functor*. There are obvious forgetful functors from **Top**, **Group**, and **Ring** to **Set** that forget about all but the underlying sets and functions.

Example 2.13. Every category $\mathcal{C}$ has an *identity functor* $Id_{\mathcal{C}}: \mathcal{C} \to \mathcal{C}$ defined by $Id_{\mathcal{C}}(A) = A$ and $Id_{\mathcal{C}}(f) = f$ on objects A and morphisms f.

Example 2.14. For a set X, let $F(X)$ denote the free group generated by X. The free group construction gives rise to a functor F: **Set** $\to$ **Group**.

Example 2.15. For a group G, let $[G, G]$ be its commutator subgroup, which is the subgroup generated by the elements $xyx^{-1}y^{-1}$ for $x, y \in G$. Then the quotient group $G_{ab} = G/[G, G]$ is an abelian group. In fact, it is the largest abelian quotient group of G in the following sense: If $\varphi: G \to H$ is a group homomorphism and if H is abelian, then there exists a unique group homomorphism $\phi: G_{ab} \to H$ such that the diagram

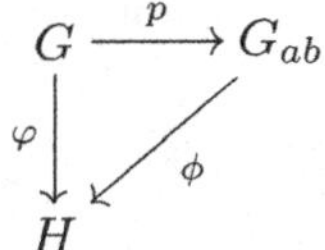

commutes. Here $p\colon G \to G_{ab}$ is the quotient map. The construction G_{ab} gives rise to a functor $(-)_{ab}\colon \mathbf{Group} \to \mathbf{Ab}$ that sends a group to its largest abelian quotient.

Example 2.16. Let A be an object in a category $\mathcal{C}$. Then there is a functor $\mathcal{C}(A,-)\colon \mathcal{C} \to \mathbf{Set}$ such that for an object B in $\mathcal{C}$, $\mathcal{C}(A,-)(B) = \mathcal{C}(A,B)$, the set of morphisms from A to B. For a morphism $g\colon B \to C$ in $\mathcal{C}$,

$$g_* = \mathcal{C}(A,-)(g)\colon \mathcal{C}(A,B) \to \mathcal{C}(A,C)$$

is the function defined by $g_*(f) = gf$ for a morphism $f\colon A \to B$.

2.2.3 *Natural transformations*

Let $F\colon \mathcal{C} \to \mathcal{D}$ and $G\colon \mathcal{C} \to \mathcal{D}$ be two functors. A *natural transformation* $\tau\colon F \to G$ is a rule that assigns to every object A in $\mathcal{C}$ a morphism $\tau_A\colon F(A) \to G(A)$ in $\mathcal{D}$. It is required to satisfy the following property: If $f\colon A \to B$ is a morphism in $\mathcal{C}$, then the square

$$\begin{array}{ccc} F(A) & \xrightarrow{\tau_A} & G(A) \\ {\scriptstyle F(f)}\downarrow & & \downarrow{\scriptstyle G(f)} \\ F(B) & \xrightarrow[\tau_B]{} & G(B) \end{array} \tag{2.18}$$

in $\mathcal{D}$ commutes. In this case, we say that τ_A is *natural in* A.

A natural transformation $\tau\colon F \to G$ is called a *natural isomorphism* if $\tau_A\colon F(A) \to G(A)$ is an isomorphism for every object A.

If $H\colon \mathcal{C} \to \mathcal{D}$ is another functor and $\nu\colon G \to H$ is a natural transformation, then we can form the natural transformation $\nu\tau\colon F \to H$ by defining

$$(\nu\tau)_A = \nu_A\tau_A\colon F(A) \to H(A)$$

for an object A in $\mathcal{C}$.

One should think of a natural transformation $\tau\colon F \to G$ as a morphism of functors. As such, it is useful to depict it using the following 2-*cell diagram*:

$$\mathcal{C} \underset{G}{\overset{F}{\rightrightarrows}} \mathcal{D}, \quad \Downarrow\tau$$

We will use this notation for natural transformations in the following examples.

Example 2.17. There is a natural transformation

$$\mathbf{Group} \underset{(-)_{ab}}{\overset{Id}{\rightrightarrows}} \mathbf{Group} \quad \Downarrow\tau$$

where Id: **Group** $\to$ **Group** is the identity functor, and $(-)_{ab}$: **Group** $\to$ **Group** is the abelianization functor discussed in Example 2.15. (Here we consider G_{ab} only as a group.) For a group G,

$$p = \tau_G \colon G \to G_{ab} = \frac{G}{[G,G]}$$

is the projection map.

Example 2.18. Let $f \colon A \to B$ be a morphism in a category $\mathcal{C}$. There is a natural transformation

$$\mathcal{C} \underset{\mathcal{C}(A,-)}{\overset{\mathcal{C}(B,-)}{\rightrightarrows}} \textbf{Set} \quad (\Downarrow f^*)$$

where $\mathcal{C}(B,-), \mathcal{C}(A,-) \colon \mathcal{C} \to$ **Set** are the morphism set functors discussed in Example 2.16. For an object C in $\mathcal{C}$, $f^*_C \colon \mathcal{C}(B,C) \to \mathcal{C}(A,C)$ is the function defined by $f^*_C(g) = gf$ for a morphism $g \colon B \to C$. In fact, it is easy to see that, given a morphism $h \colon C \to D$ in $\mathcal{C}$, the diagram

$$\begin{array}{ccc} \mathcal{C}(B,C) & \xrightarrow{f^*_C} & \mathcal{C}(A,C) \\ \downarrow{\scriptstyle \mathcal{C}(B,-)(h)} & & \downarrow{\scriptstyle \mathcal{C}(A,-)(h)} \\ \mathcal{C}(B,D) & \xrightarrow[f^*_D]{} & \mathcal{C}(A,D) \end{array}$$

in **Set** commutes because composition of functions is associative.

2.3 Comonads and Coalgebras

A usual monoid consists of a set G, an associative multiplication $m \colon G \times G \to G$, and an identity element e for the multiplication. The associativity and identity axioms can be expressed as the commutative diagrams

$$\begin{array}{ccc} G \times G \times G & \xrightarrow{m \times Id} & G \times G \\ \downarrow{\scriptstyle Id \times m} & & \downarrow{\scriptstyle m} \\ G \times G & \xrightarrow{m} & G \end{array} \tag{2.19}$$

and

$$\begin{array}{ccccc} G & \xrightarrow{(e,Id)} & G \times G & \xleftarrow{(Id,e)} & G \\ & \searrow\!\!\!\!= & \downarrow{\scriptstyle m} & =\!\!\!\!\swarrow & \\ & & G & & \end{array} , \tag{2.20}$$

respectively.

The corresponding categorical notion is called a *monad.* Thus, in a monad, G is replaced by a functor, and m and e are replaced by some natural transformations. The axioms for a monad are expressed as commutative diagrams of functors and natural transformations similar to (2.19) and (2.20) above. If one invert the arrows in these diagrams, one gets what is called a comonad, which we now introduce. The references for comonads are [Mac Lane (1998); Weibel (1994)].

2.3.1 *Comonads*

Let $\mathcal{C}$ be a category.

Definition 2.19. A *comonad on* $\mathcal{C}$ is a triple $T = (T, \delta, \varepsilon)$ consisting of a functor $T\colon \mathcal{C} \to \mathcal{C}$ and natural transformations $\delta\colon T \to T^2$ and $\varepsilon\colon T \to Id$. It is required that the diagrams

$$\begin{array}{ccc} T & \xrightarrow{\delta} & T^2 \\ {\scriptstyle\delta}\downarrow & & \downarrow{\scriptstyle T\delta} \\ T^2 & \xrightarrow{\delta T} & T^3 \end{array} \tag{2.21}$$

and

$$\begin{array}{ccccc} & & T & & \\ & \swarrow\!\!= & \downarrow{\scriptstyle\delta} & =\!\!\searrow & \\ T & \xleftarrow{\varepsilon T} & T^2 & \xrightarrow{T\varepsilon} & T \end{array} \tag{2.22}$$

be commutative.

The natural transformations δ and ε are called the *comultiplication* and the *counit*, respectively. Thus, a comonad is a functor T together with a coassociative comultiplication and a counit that is compatible with the comultiplication.

Remark 2.20. In the literature (e.g., [Weibel (1994)]), a comonad is sometimes called a *cotriple.*

In practice, one is usually not so much interested in a comonad in and of itself. Rather, one is interested in certain objects, called *coalgebras*, that are associated to a comonad.

2.3.2 *Coalgebras*

Let $T = (T, \delta, \varepsilon)$ be a comonad on a category $\mathcal{C}$.

Definition 2.21. A T*-coalgebra* consists of an object A in $\mathcal{C}$ and a morphism

$\rho\colon A \to T(A)$ in $\mathcal{C}$ such that the diagrams

$$\begin{array}{ccc} A & \xrightarrow{\rho} & T(A) \\ & \diagdown\!\!\diagdown & \downarrow{\scriptstyle \varepsilon_A} \\ & & A \end{array} \tag{2.23}$$

and

$$\begin{array}{ccc} A & \xrightarrow{\rho} & T(A) \\ {\scriptstyle \rho}\downarrow & & \downarrow{\scriptstyle T\rho} \\ T(A) & \xrightarrow{\delta_A} & T^2(A). \end{array} \tag{2.24}$$

commute.

The morphism ρ is called the *structure map* of the T-coalgebra A. The two commutative diagrams above say that the structure map is counital and is coassociative with respect to the comultiplication in T. We will denote a T-coalgebra by (A, ρ_A) or simply A, if the structure map is understood.

If (A, ρ_A) and (B, ρ_B) are two T-coalgebras, then a *morphism* $f\colon (A, \rho_A) \to (B, \rho_B)$ of T-coalgebras is a morphism $f\colon A \to B$ in $\mathcal{C}$ such that the square

$$\begin{array}{ccc} A & \xrightarrow{\rho_A} & T(A) \\ {\scriptstyle f}\downarrow & & \downarrow{\scriptstyle Tf} \\ B & \xrightarrow{\rho_B} & T(B) \end{array} \tag{2.25}$$

commutes.

The collection of T-coalgebras and their morphisms form a category, denoted by $\mathcal{C}^T$. There is a forgetful functor $U\colon \mathcal{C}^T \to \mathcal{C}$ that forgets about the structure maps on objects. If (A, ρ_A) is a T-coalgebra, then we will simply write A for the underlying object $U(A)$ in $\mathcal{C}$. The same convention goes for morphisms of T-coalgebras.

There is a closely related functor that goes in the other direction.

Theorem 2.22. *Let $T = (T, \delta, \varepsilon)$ be a comonad on a category $\mathcal{C}$. Then there is a functor $F\colon \mathcal{C} \to \mathcal{C}^T$ defined by $F(A) = (T(A), \delta_A)$ and $F(f) = T(f)$ on objects A and morphisms f in $\mathcal{C}$.*

Proof. First we need to check that $T(A)$ is a T-coalgebra with structure map $\delta_A\colon T(A) \to T^2(A)$. In other words, we need to check that the diagrams (2.23) and (2.24) are commutative when A and ρ are replaced by $T(A)$ and δ_A, respectively. The resulting diagrams are

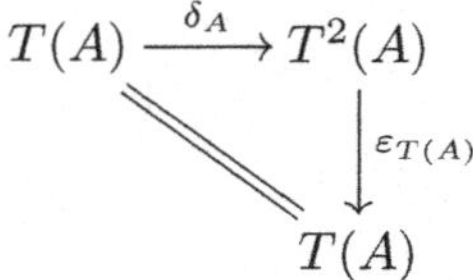

and

$$\begin{array}{ccc} T(A) & \xrightarrow{\delta_A} & T^2(A) \\ {\scriptstyle \delta_A}\downarrow & & \downarrow{\scriptstyle T(\delta_A)} \\ T^2(A) & \xrightarrow{\delta_{T(A)}} & T^3(A). \end{array}$$

The first diagram is the left commutative triangle in (2.22) applied to A, so it is commutative. Likewise, the second diagram is the commutative diagram (2.21) applied to A, so it is commutative as well.

Next we need to check that, if $f\colon A \to B$ is a morphism in $\mathcal{C}$, then $T(f)\colon (T(A), \delta_A) \to (T(B), \delta_B)$ is a morphism of T-coalgebras. This is equivalent to checking the commutativity of the square

$$\begin{array}{ccc} T(A) & \xrightarrow{\delta_A} & T^2(A) \\ {\scriptstyle T(f)}\downarrow & & \downarrow{\scriptstyle T^2(f)} \\ T(B) & \xrightarrow{\delta_B} & T^2(B), \end{array}$$

which we know is true since $\delta\colon T \to T^2$ is a natural transformation.

Finally, we need to show that F preserves the identity morphisms and compositions. For any object A in $\mathcal{C}$, we have

$$F(1_A) = T(1_A) = 1_{T(A)} = 1_{F(A)},$$

since T is a functor. Likewise, F preserves compositions of morphisms because T does. $\square$

Definition 2.23. The T-coalgebra $(T(A), \delta_A)$ (Theorem 2.22) is called the *free T-coalgebra on A.*

The object $T(A)$ is called the free T-coalgebra on A because, in some sense, it is the closest T-coalgebra to A from the left. The following result makes this statement precise.

Theorem 2.24. *Let $T = (T, \delta, \varepsilon)$ be a comonad on a category $\mathcal{C}$, and let A be an object in $\mathcal{C}$ and (B, ρ_B) be a T-coalgebra. Then for every morphism $f\colon B \to A$ in $\mathcal{C}$, there exists a unique morphism $\varphi\colon B \to T(A)$ of T-coalgebras such that the diagram*

$$\begin{array}{ccc} & & T(A) \\ & {\scriptstyle \varphi}\nearrow & \downarrow{\scriptstyle \varepsilon_A} \\ B & \xrightarrow{f} & A \end{array}$$

commutes in $\mathcal{C}$.

Proof. It suffices to show that the map

$$\eta \colon \mathcal{C}(B, A) \to \mathcal{C}^T(B, T(A))$$

defined by $\eta(f) = T(f)\rho_B$ is a bijection and has the property that

$$\varepsilon_A \eta(f) = f \tag{2.26}$$

for any morphism $f \colon B \to A$ in $\mathcal{C}$. If these two assertions are true, then we will simply take φ to be $\eta(f)$.

First we prove (2.26). Observe that the diagram

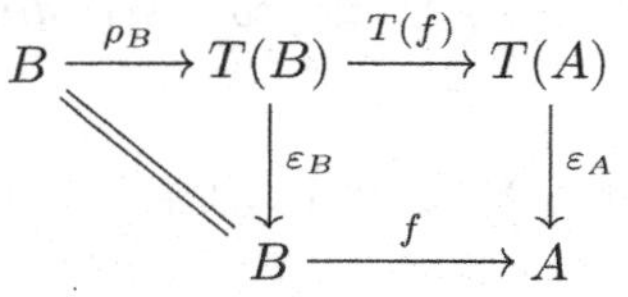

$$\begin{array}{ccccc} B & \xrightarrow{\rho_B} & T(B) & \xrightarrow{T(f)} & T(A) \\ & \diagdown\!\!\diagdown & \downarrow{\scriptstyle \varepsilon_B} & & \downarrow{\scriptstyle \varepsilon_A} \\ & & B & \xrightarrow{f} & A \end{array}$$

is commutative. Indeed, the left triangle is commutative because of the axiom (2.23) of a T-coalgebra. The right square is commutative because $\varepsilon \colon T \to Id$ is a natural transformation. This proves (2.26).

To show that η is a bijection, consider the map $\theta \colon \mathcal{C}^T(B, T(A)) \to \mathcal{C}(B, A)$ defined by $\theta(\phi) = \varepsilon_A \phi$. It suffices to show that η and θ are the inverses of each other. We have already shown that $\theta(\eta(f)) = \varepsilon_A \eta(f) = f$, which is just (2.26).

For the converse, pick a morphism $\phi \colon B \to T(A)$ of T-coalgebras. We need to show that

$$\eta(\theta(\phi)) = T(\varepsilon_A \phi)\rho_B = \phi. \tag{2.27}$$

Since T is a functor, we have $T(\varepsilon_A \phi) = T(\varepsilon_A)T(\phi)$. Consider the diagram

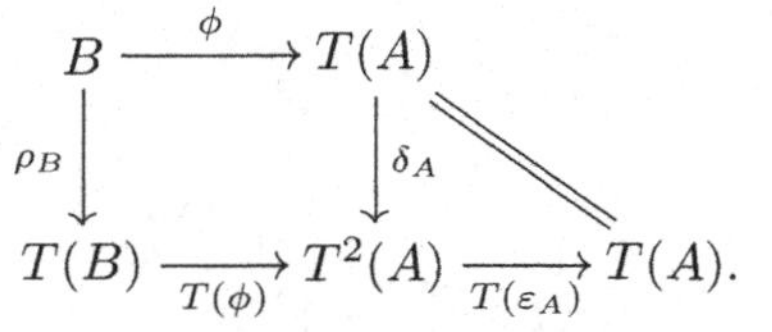

$$\begin{array}{ccccc} B & \xrightarrow{\phi} & T(A) & & \\ \downarrow{\scriptstyle \rho_B} & & \downarrow{\scriptstyle \delta_A} & \diagdown\!\!\diagdown & \\ T(B) & \xrightarrow[T(\phi)]{} & T^2(A) & \xrightarrow[T(\varepsilon_A)]{} & T(A). \end{array}$$

We claim that it is commutative, which will imply (2.27). The left square is commutative because of the axiom (2.25) for a T-coalgebra morphism. The right triangle is the right commutative triangle in (2.22) applied to A, so it is commutative. □

2.4 λ-Rings as Λ-Coalgebras

In this section, we discuss the comonad Λ from the universal λ-ring construction whose coalgebras are exactly the λ-rings (Theorem 2.26). Recall that **Ring** denotes the category of commutative rings.

Theorem 2.25. *There is a comonad $(\Lambda, \lambda, \varepsilon)$ on* **Ring** *such that, for a ring R, $\Lambda(R)$ is the universal λ-ring of R (Theorem 2.6), and*

$$\lambda_R = \lambda_t \colon \Lambda(R) \to \Lambda^2(R)$$

is the map

$$\lambda_R\Big(1+\sum_{n\geq 1} a_n t^n\Big) = 1 + \sum_{m\geq 1}\Big(1+\sum_{n\geq 1} P_{n,m}(a_1,\ldots,a_{nm})t^n\Big)t^m.$$

The map $\varepsilon_R\colon \Lambda(R) \to R$ *is defined by*

$$\varepsilon_R\Big(1+\sum_{n\geq 1} a_n t^n\Big) = a_1.$$

For a ring homomorphism $\phi\colon R \to S$, *the corresponding map* $\Lambda(\phi)\colon \Lambda(R) \to \Lambda(S)$ *is defined by*

$$\Lambda(\phi)\Big(1+\sum_{n\geq 1} a_n t^n\Big) = 1 + \sum_{n\geq 1} \phi(a_n)t^n. \tag{2.28}$$

Proof. It is immediate from the definition that Λ preserves identity morphisms and compositions, so it is a functor on the category of rings. Since $\Lambda(R)$ is always a λ-ring (Theorem 2.6), the map $\lambda_R = \lambda_t\colon \Lambda(R) \to \Lambda^2(R)$ is actually a λ-homomorphism (Proposition 2.8). In particular, it is a ring homomorphism. The fact that $\lambda\colon \Lambda \to \Lambda^2$ is a natural transformation (i.e., the commutativity of the square (2.18)) is again immediate from the definition.

The map$\varepsilon_R\colon \Lambda(R) \to R$ is a ring homomorphism because $(1+at)\times_\Lambda(1+bt) = 1+abt$ in $\Lambda(R)$. It follows as above that $\varepsilon\colon \Lambda \to Id$ is a natural transformation.

It remains to show the commutativity of the diagrams (2.21) and (2.22) in the present case. As in the proofs of Theorems 2.5 and 2.6, for a ring R, showing that the square

$$\begin{array}{ccc} \Lambda(R) & \xrightarrow{\lambda_R} & \Lambda^2(R) \\ {\scriptstyle\lambda_R}\big\downarrow & & \big\downarrow{\scriptstyle\Lambda(\lambda_R)} \\ \Lambda^2(R) & \xrightarrow{\lambda_{\Lambda(R)}} & \Lambda^3(R) \end{array} \tag{2.29}$$

is commutative is equivalent to proving certain identity involving the elementary symmetric functions. We leave it to the reader as an exercise.

The proof of the commutativity of (2.22) in the present case is by direct inspection. Indeed, suppose that $f = 1+\sum_{n\geq 1} a_n t^n$ is an element in $\Lambda(R)$. Then we have

$$\begin{aligned} \Lambda\varepsilon_R\left(\lambda_R(f)\right) &= 1+\sum_{m\geq 1}\left(\varepsilon_R\lambda^m(f)\right)t^m \\ &= 1+\sum_{m\geq 1} P_{1,m}(a_1,\ldots,a_m)t^m \\ &= 1+\sum_{m\geq 1} a_m t^m \\ &= f. \end{aligned}$$

In the third equality we used (1.6). Similarly, we have

$$\varepsilon_{\Lambda(R)}\left(\lambda_R(f)\right) = \lambda^1(f) = f.$$

We have shown that $\varepsilon_{\Lambda(R)}\lambda_R = Id = (\Lambda\varepsilon_R)\lambda_R$, as desired. □

In Corollary 4.22 below, we will give a characterization of the functor Λ: **Ring** $\to$ **Ring** in terms of some natural transformations $z_n\colon \Lambda \to Id$ for $n \geq 1$.

Since Λ is a comonad, it has a corresponding category of Λ-coalgebras (Definition 2.21). We now identify these coalgebras with λ-rings.

Theorem 2.26. *For the comonad $(\Lambda, \lambda, \varepsilon)$ in* **Ring**, *the Λ-coalgebras are λ-rings, and vice versa.*

Proof. A Λ-coalgebra is a ring R together with a ring homomorphism $\lambda_t\colon R \to \Lambda(R)$ such that

$$\varepsilon_R\lambda_t = Id_R, \tag{2.30a}$$

$$\lambda_R\lambda_t = \Lambda(\lambda_t)\lambda_t. \tag{2.30b}$$

Given such a Λ-coalgebra, we define the λ-operations in R as the components of the structure map λ_t:

$$\lambda_t(a) = 1 + \sum_{n\geq 1} \lambda^n(a)t^n \in \Lambda(R)$$

for $a \in R$. Thus, $\lambda^0(a) = 1$ by definition. The condition (2.30a) is equivalent to $\lambda^1 = Id$. Since λ_t preserves the multiplicative identity, we have $\lambda_t(1) = 1 + t$. So axioms (1) to (3) in Definition 1.10 are satisfied.

The additivity of the map λ_t is equivalent to axiom (4) in Definition 1.10. Indeed, on the one hand, we have

$$\begin{aligned}\lambda_t(a) +_\Lambda \lambda_t(b) &= \left(1 + \sum_{n\geq 1} \lambda^n(a)t^n\right)\left(1 + \sum_{n\geq 1} \lambda^n(b)t^n\right)\\ &= 1 + \sum_{n\geq 1}\left(\sum_{i=0}^{n} \lambda^i(a)\lambda^{n-i}(b)\right)t^n.\end{aligned}$$

On the other hand, the coefficient of t^n in $\lambda_t(a+b)$ is $\lambda^n(a+b)$. Similarly, the multiplicativity of λ_t is equivalent to axiom (5) of a λ-ring.

Finally, (2.30b) is equivalent to axiom (6) in Definition 1.10. In fact, we have

$$\begin{aligned}\Lambda(\lambda_t)(\lambda_t(a)) &= 1 + \sum_{m\geq 1} \left(\lambda_t(\lambda^m(a))\right)t^m\\ &= 1 + \sum_{m\geq 1}\left(1 + \sum_{n\geq 1} \lambda^n(\lambda^m(a))t^n\right)t^m.\end{aligned}$$

Likewise, we have

$$\begin{aligned}\lambda_R\lambda_t(a) &= 1 + \sum_{m\geq 1} \lambda^m\left(1 + \sum_{n\geq 1} \lambda^n(a)t^n\right)t^m\\ &= 1 + \sum_{m\geq 1}\left(1 + \sum_{n\geq 1} P_{n,m}(\lambda^1(a), \ldots, \lambda^{nm}(a))t^n\right)t^m.\end{aligned}$$

Comparing the coefficients of t^m, we see that (2.30b) is equivalent to axiom (6).

For the converse, simply reverse the arguments above. □

Example 2.27. For the λ-ring $\mathbf{Z}$, its structure map as a Λ-coalgebra is $\lambda_t \colon \mathbf{Z} \to \Lambda(\mathbf{Z})$, where

$$\begin{aligned}\lambda_t(n) &= (1+t)^n \\ &= 1 + nt + \binom{n}{2}t^2 + \cdots + \binom{n}{n-1}t^{n-1} + t^n\end{aligned}$$

for $0 \leq n \in \mathbf{Z}$. If $n = -m$ where $m > 0$, then

$$\begin{aligned}\lambda_t(n) &= (1+t)^{-m} \\ &= 1 - mt + \frac{m(m+1)}{2}t^2 + \cdots .\end{aligned}$$

Corollary 2.28. *Let R be a ring and S be a λ-ring. Then there is a one-to-one correspondence between ring homomorphisms $f \colon S \to R$ and λ-homomorphisms $\varphi \colon S \to \Lambda(R)$.*

Proof. Apply Theorem 2.24 to the comonad $(\Lambda, \lambda, \varepsilon)$, and identify λ-rings with Λ-coalgebras. It is immediate from the definition that a morphism of Λ-coalgebras is equivalent to a λ-homomorphism. □

2.5 Exercises

(1) Prove (2.16).

(2) Prove the commutativity of the diagram (2.29).

(3) Let R be a λ-ring with structure map $\lambda_t \colon R \to \Lambda(R)$. Describe the structure map λ_t for the polynomial λ-ring $R[x_1, \ldots, x_n]$ (Corollary 1.35) in which each variable x_i has dimension 1.

(4) Prove that the polynomial $i_n(x_1, \ldots, x_n)$ (2.2) is homogeneous of degree n if x_j is given degree j.

(5) In Corollary 2.28, describe explicitly the map φ in terms of f.

(6) A $\mathbf{Q}$-*algebra* is a ring R that comes equipped with an injective ring homomorphism $\varphi \colon \mathbf{Q} \to R$. Let R be a λ-ring that is also a $\mathbf{Q}$-algebra. Show that the universal λ-ring $\Lambda(R)$ has a $\mathbf{Q}$-algebra structure that extends the one on R.

Chapter 3

Adams Operations

In this chapter we discuss Adams operations, which were first introduced by Adams [Adams (1962)] in his solution to a problem about the maximum number of linearly independent vector fields on spheres. It was soon recognized that Adams operations were useful in many other settings. Recall that λ-operations are neither additive nor multiplicative. Moreover, the composition of two λ-operations $\lambda^n(\lambda^m(x))$ is given by a rather complicated formula in terms of the individual λ-operations $\lambda^i(x)$.

Adams operations $\psi^n\colon R \to R$ for $n \geq 1$ exist in any λ-ring R. They are much easier to work with when it comes to actual computation. In fact, each ψ^n is a λ-homomorphism on the given λ-ring, so, in particular, they are both additive and multiplicative. The composition $\psi^n\psi^m$ is equal to ψ^{nm}, so the Adams operations are determined by the ψ^p for p prime. Moreover, ψ^p is an integral lift of the Frobenius map modulo p (Theorem 3.7). Every λ-homomorphism must be compatible with all the Adams operations. In other words, Adams operations are natural operations on λ-rings (Theorem 3.6).

Adams operations can be expressed in terms of the λ-operations in two ways. First, there is a closed formula expressing ψ^n in terms of the λ-operations. Since Adams operations are natural operations on λ-rings, by Theorem 1.42 each ψ^n must be a polynomial with integer coefficients in the λ-operations. In fact, ψ^n can be written as a polynomial

$$\psi^n(x) = Q_n(\lambda^1(x), \ldots, \lambda^n(x))$$

in $\lambda^1, \ldots, \lambda^n$ (Theorem 3.9). Second, there is an inductive formula, called the *Newton Formula* (Theorem 3.10), relating $\psi^1, \ldots, \psi^n$ and $\lambda^1, \ldots, \lambda^n$. This is useful for induction type arguments. All of these are discussed in §3.1.

In §3.2 and §3.3 we discuss γ-operations $\gamma^n\colon R \to R$ in a λ-ring R following [Atiyah and Tall (1969)]. Every γ-operation γ^n is a $\mathbf{Z}$-linear combination of $\lambda^1, \ldots, \lambda^n$, and vice versa (Corollary 3.20). The operations γ^n satisfy some axioms similar to those for λ-rings (Corollary 3.24). For an augmented λ-ring, there is an associated γ-filtration (Proposition 3.27). The Adams operation ψ^k acts as k^n on the nth quotient of the γ-filtration (Theorem 3.41).

Sometimes one encounters a ring R that is equipped with certain mutually commuting ring endomorphisms ψ^k that behave like the Adams operations in a λ-ring.

This is called a *ψ-ring*. It is useful to know whether a ψ-ring actually determines a λ-ring structure. When R is $\mathbf{Z}$-torsion-free, a result due to Knutson [Knutson (1973)] says that these operations ψ^k do, in fact, determine a λ-ring structure on $R \otimes \mathbf{Q}$. These λ-operations are obtained by inverting the Newton Formulas. This result of Knutson is discussed in §3.4.

In §3.5 we describe the comonad Ψ on the category of rings whose coalgebras are exactly the ψ-rings (Theorem 3.53).

One might wonder if the above process of going from Adams-type operations to λ-operations can be improved so that the λ-operations are defined on R itself. Wilkerson's Theorem [Wilkerson (1982)] says that this is indeed possible, provided that one assumes $\psi^p(a) \equiv a^p \pmod{pR}$ as well for $a \in R$ and p prime. Moreover, in this case there is a *unique* λ-ring structure on R that has the given ψ^n as Adams operations. Thus, in the absence of $\mathbf{Z}$-torsion, Adams operations uniquely determine the λ-operations. Wilkerson's Theorem 3.54 is the subject of §3.6.

In §3.7 we apply Wilkerson's Theorem to localizations and completions of λ-rings. It is shown that λ-rings are closed under most reasonable localizations and completions. In particular, the p-local integers $\mathbf{Z}_{(p)}$ and the p-adic integers $\hat{\mathbf{Z}}_p$ are both λ-rings with $\psi^k = Id$ for all k.

3.1 Adams Operations

In this section we define Adams operations in a λ-ring and establish some of their basic properties, most of which can be found in [Adams (1962); Atiyah and Tall (1969); Knutson (1973)].

Throughout this section, let R be a λ-ring (Definition 1.10).

Definition 3.1. For each $n \geq 1$ define the *nth Adams operation* $\psi^n \colon R \to R$ by the generating function

$$\psi_{-t}(x) = -t\frac{d}{dt}(\log \lambda_t(x)) = \frac{-t\lambda_t(x)'}{\lambda_t(x)} \tag{3.1}$$

for $x \in R$, where $\psi_t(x) = \sum_{k\geq 1} \psi^k(x)t^k$. Here log means natural logarithm and $\lambda_t(x)'$ denotes the derivative of the formal power series $\lambda_t(x)$ (1.7) with respect to t.

In other words, $\psi^k(x)$ is the coefficient of $(-t)^k$ in (3.1).

Remark 3.2. Adams operations can be defined as above for pre-λ-rings (Definition 1.10), since the map λ_t is defined. Some of the properties of Adams operations in a λ-ring also hold in a pre-λ-ring.

Example 3.3. We compute the first few $\psi^n(x)$ here. First we have

$$\lambda_t(x)^{-1} = 1 - xt + (x^2 - \lambda^2(x))t^2 + (-x^3 + 2x\lambda^2(x) - \lambda^3(x))t^3 + \cdots .$$

Putting this back into $\psi_{-t}(x)$, we have

$$\begin{aligned}\psi_{-t}(x) &= -t\lambda_t(x)'\left(\lambda_t(x)\right)^{-1} \\ &= -t\left(x + 2\lambda^2(x)t + 3\lambda^3(x)t^2 + 4\lambda^4(x)t^3 + \cdots\right) \times \\ &\quad \left(1 - xt + (x^2 - \lambda^2(x))t^2 + (-x^3 + 2x\lambda^2(x) - \lambda^3(x))t^3 + \cdots\right) \\ &= -xt + \left(x^2 - 2\lambda^2(x)\right)t^2 + \left(-x^3 + 3x\lambda^2(x) - 3\lambda^3(x)\right)t^3 + \\ &\quad \left(x^4 - 4x^2\lambda^2(x) + 4x\lambda^3(x) + 2(\lambda^2(x))^2 - 4\lambda^4(x)\right)t^4 + \cdots .\end{aligned}$$

Thus, comparing the coefficients of $(-t)^n$, we obtain:

$$\begin{aligned}\psi^1(x) &= x, \\ \psi^2(x) &= x^2 - 2\lambda^2(x), \\ \psi^3(x) &= x^3 - 3x\lambda^2(x) + 3\lambda^3(x), \\ \psi^4(x) &= x^4 - 4x^2\lambda^2(x) + 4x\lambda^3(x) + 2\left(\lambda^2(x)\right)^2 - 4\lambda^4(x).\end{aligned}$$

Moreover, if we set $\deg(\lambda^n(x)) = n$, then $\psi^n(x)$ is a homogeneous polynomial with integer coefficients of degree n in $\lambda^1(x), \ldots, \lambda^n(x)$. This polynomial will be described in Theorem 3.9.

Example 3.4. Here we compute the Adams operations in the λ-ring $\mathbf{Z}$, in which $\lambda_t(n) = (1+t)^n$ for an integer n. So we have

$$\psi_{-t}(n) = -t\frac{d}{dt}\left(\log(1+t)^n\right) = \frac{-nt}{1+t},$$

which implies that

$$\psi_t(n) = \frac{nt}{1-t} = n(t + t^2 + t^3 + \cdots).$$

Therefore, we have $\psi^k(n) = n$ for all $k \geq 1$ and $n \in \mathbf{Z}$. In other words, in the λ-ring $\mathbf{Z}$, all the Adams operations are equal to the identity. As we will see shortly, ψ^k is actually a ring homomorphism (Theorem 3.6), so it must preserve the integers.

Proposition 3.5. *Let R be a λ-ring. Then each Adams operation ψ^n is additive and preserves the additive zero. Moreover, if $a \in R$ has dimension 1, then $\psi^n(a) = a^n$. In particular, we have*

$$\psi^n(a_1 + \cdots + a_m) = a_1^n + \cdots + a_m^n$$

for 1-dimensional elements $a_1, \ldots, a_m \in R$.

Proof. To show that ψ^n is additive, pick elements a and b in R. Since the map λ_t is additive (1.8) and $+_\Lambda$ (2.4) is multiplication of strict power series, we have

$$\begin{aligned}\psi_{-t}(a+b) &= -t\frac{d}{dt}\left(\log\lambda_t(a+b)\right) \\ &= -t\frac{d}{dt}\left(\log(\lambda_t(a)\lambda_t(b))\right) \\ &= -t\frac{d}{dt}\left(\log\lambda_t(a) + \log\lambda_t(b)\right) \\ &= \psi_{-t}(a) + \psi_{-t}(b).\end{aligned}$$

Equating the coefficients of $(-t)^n$ on both sides of this equality, we see that ψ^n is additive. Each ψ^n preserves the additive zero because

$$\begin{aligned}\psi_{-t}(0) &= -t\frac{d}{dt}(\log \lambda_t(0))\\ &= -t\frac{d}{dt}(\log 1)\\ &= 0.\end{aligned}$$

For the second assertion, let $a \in R$ be of dimension 1. Then we have $\lambda_t(a) = 1 + at$, which implies that

$$\begin{aligned}\psi_{-t}(a) &= -t\frac{d}{dt}(\log(1+at))\\ &= \frac{-at}{1+at}\\ &= -at + a^2t^2 - a^3t^3 + \cdots .\end{aligned}$$

The coefficient of $(-t)^n$ gives $\psi^n(a) = a^n$. This last assertion follows immediately from the first two. $\square$

Theorem 3.6. *Each Adams operation is a natural operation on λ-rings. Moreover, for a λ-ring R, every Adams operation $\psi^n\colon R \to R$ is a λ-homomorphism.*

Proof. The first assertion that ψ^n is a natural operation on λ-rings means that, if $\varphi\colon R \to S$ is a λ-homomorphism, then the square

$$\begin{array}{ccc} R & \xrightarrow{\varphi} & S \\ {\scriptstyle\psi^n}\downarrow & & \downarrow{\scriptstyle\psi^n} \\ R & \xrightarrow{\varphi} & S \end{array}$$

is commutative, i.e., $\psi^n\varphi = \varphi\psi^n$. This is equivalent to $\psi_{-t}(\varphi(x)) = \varphi(\psi_{-t}(x))$ for $x \in R$. Since φ is ring homomorphism and commutes with the λ-operations, we have

$$\begin{aligned}\psi_{-t}(\varphi(x)) &= -t\frac{d}{dt}\left(\log \lambda_t(\varphi(x))\right)\\ &= -t\frac{d}{dt}\left(\log \varphi(\lambda_t(x))\right)\\ &= \frac{-t\varphi(\lambda_t(x))'}{\varphi(\lambda_t(x))}\\ &= \varphi\left(\frac{-t\lambda_t(x)'}{\lambda_t(x)}\right)\\ &= \varphi(\psi_{-t}(x)),\end{aligned}$$

as desired.

To show that ψ^n is a λ-homomorphism, note that we already know that ψ^n is additive. To show that it is multiplicative, by the Verification Principle (Theorem 1.43), it suffices to consider finite sums of 1-dimensional elements, i.e.,

$$\psi^n\Bigl(\sum_i a_i \times \sum_j b_j\Bigr) = \psi^n\Bigl(\sum_i a_i\Bigr)\psi^n\Bigl(\sum_j b_j\Bigr),$$

where the a_i and b_j all have dimension 1. Since ψ^n is additive (Proposition 3.5), it suffices to show that $\psi^n(ab) = \psi^n(a)\psi^n(b)$ when a and b both have dimension 1. Recall that in a λ-ring, the product of two 1-dimensional elements is again 1-dimensional (Proposition 1.13). So we have

$$\psi^n(ab) = (ab)^n = a^n b^n = \psi^n(a)\psi^n(b),$$

as desired. Moreover, ψ^n preserves the multiplicative identity because

$$\begin{aligned}\psi_{-t}(1) &= -t\frac{d}{dt}(\log \lambda_t(1)) \\ &= -t\frac{d}{dt}(\log(1+t)) \\ &= \frac{-t}{1+t},\end{aligned}$$

which implies that

$$\begin{aligned}\psi_t(1) &= \frac{t}{1-t} \\ &= t + t^2 + t^3 + \cdots .\end{aligned}$$

This shows that $\psi^n(1) = 1$.

Finally, to show that $\psi^n\lambda^m = \lambda^m\psi^n$, by the Verification Principle (Theorem 1.43) again, it suffices to consider finite sums of 1-dimensional elements. So let $a_1, \ldots, a_r$ be 1-dimensional elements. Then we have

$$\begin{aligned}\psi^n(\lambda^m(a_1 + \cdots + a_r)) &= \psi^n\Bigl(\sum_{1\le i_1<\cdots<i_m\le r} a_{i_1}\cdots a_{i_m}\Bigr) \\ &= \sum_{1\le i_1<\cdots<i_m\le r} \psi^n(a_{i_1})\cdots\psi^n(a_{i_m}) \\ &= \sum_{1\le i_1<\cdots<i_m\le r} a_{i_1}^n\cdots a_{i_m}^n \\ &= \lambda^m(a_1^n + \cdots + a_r^n) \\ &= \lambda^m(\psi^n(a_1 + \cdots + a_r)).\end{aligned}$$

In the fourth equality above, we used the fact that each a_i^n is 1-dimensional. □

It turns out that the Adams operations commute with each other. Moreover, ψ^n is uniquely determined by ψ^p, where p runs through the prime factors of n. When p is a prime, ψ^p is an integral lift of the Frobenius (i.e., pth power) map modulo p.

Theorem 3.7. *The following statements hold in a λ-ring R.*

(1) *For integers $m, n \geq 1$, we have*

$$\psi^m \psi^n = \psi^{mn} = \psi^n \psi^m. \tag{3.2}$$

(2) *If n has prime factorization $n = p_1^{e_1} \cdots p_k^{e_k}$, where $p_1, \ldots, p_k$ are primes and each $e_i > 0$, then*

$$\psi^n = (\psi^{p_1})^{e_1} \cdots (\psi^{p_k})^{e_k}. \tag{3.3}$$

(3) *If p is a prime and $a \in R$, then*

$$\psi^{p^n}(a) \equiv a^{p^n} \pmod{pR} \tag{3.4}$$

for all $n \geq 1$. In particular, we have

$$\psi^p(a) \equiv a^p \pmod{pR}$$

when $n = 1$.

Proof. To prove (3.2), by The Verification Principle (Theorem 1.43) it suffices to show that $\psi^m \psi^n = \psi^{mn}$ when applied to a finite sum of 1-dimensional elements $a_1, \ldots, a_r$. In this case, we have

$$\begin{aligned}\psi^m(\psi^n(a_1 + \cdots + a_r)) &= \psi^m(a_1^n + \cdots + a_r^n)\\ &= a_1^{mn} + \cdots + a_r^{mn}\\ &= \psi^{mn}(a_1 + \cdots + a_r),\end{aligned}$$

as desired. In the first and the last equalities, we used Proposition 3.5. In the second equality, we used the fact that each a_i^n is still 1-dimensional (Proposition 1.13).

The condition (3.3) follows from repeated applications of (3.2).

To prove (3.4), we again assume that a is a sum of 1-dimensional elements $a_1, \ldots, a_r$. We have

$$\begin{aligned}\psi^{p^n}(a_1 + \cdots + a_r) &= a_1^{p^n} + \cdots + a_r^{p^n}\\ &\equiv (a_1 + \cdots + a_r)^{p^n} \pmod{pR},\end{aligned}$$

since the pth power map is linear modulo p. □

Since each Adams operation ψ^n is a natural operation on λ-rings (Theorem 3.6), it follows from Theorem 1.42 that ψ^n is a polynomial with integer coefficients in the λ-operations. In fact, as it was pointed out in Example 3.3, ψ^n is a homogeneous polynomial of degree n in $\lambda^1, \ldots, \lambda^n$, where λ^i is given degree i. We now give a more explicit description of this polynomial.

Let $x_1, \ldots, x_n$ be variables and $s_1, \ldots, s_n$ be their elementary symmetric functions (Example 1.1). Since the kth power sum polynomial $x_1^k + \cdots + x_n^k$ is symmetric, by Theorem 1.4 there exists a unique polynomial Q_k in n variables with integer coefficients such that

$$x_1^k + \cdots + x_n^k = Q_k(s_1, \ldots, s_n). \tag{3.5}$$

For example, we have

$$Q_2(s_1, s_2) = s_1^2 - 2s_2$$

by Example 1.6. Likewise, a little computation shows that

$$Q_3(s_1, s_2, s_3) = s_1^3 - 3s_1s_2 + 3s_3$$

and

$$Q_4(s_1, s_2, s_3, s_4) = s_1^4 - 4s_1^2s_2 + 2s_2^2 + 4s_1s_3 - 4s_4.$$

Lemma 3.8. *Consider the polynomial $f(t) = 1 + x_1t + x_2t^2 + \cdots + x_nt^n$. Then the equality*

$$-t\frac{d}{dt}(\log f(t)) = \sum_{k\geq 1} Q_k(x_1, \ldots, x_n)(-t)^k$$

holds.

Proof. It suffices to consider the universal case when the x_i are algebraically independent. So it suffices to consider the case $x_i = s_i(y_1, \ldots, y_n)$, the ith elementary symmetric function on $y_1, \ldots, y_n$, for $1 \leq i \leq n$, where $y_1, \ldots, y_n$ are independent variables. In this case, we have $f(t) = \prod_{i=1}^n (1 + y_it)$, which implies that

$$\begin{aligned}
-t\frac{d}{dt}(\log f(t)) &= -t\frac{d}{dt}\left(\sum_{i=1}^n \log(1 + y_it)\right) \\
&= \sum_{i=1}^n \frac{-y_it}{1 + y_it} \\
&= \sum_{i=1}^n \sum_{k\geq 1} y_i^k(-t)^k \\
&= \sum_{k\geq 1} (y_1^k + \cdots + y_n^k)(-t)^k \\
&= \sum_{k\geq 1} Q_k(x_1, \ldots, x_n)(-t)^k.
\end{aligned}$$

This proves the Lemma. □

Theorem 3.9. *In any λ-ring R, we have*

$$\psi^n(x) = Q_n(\lambda^1(x), \ldots, \lambda^n(x))$$

for $x \in R$ and $n \geq 1$, where the polynomial Q_n with integer coefficients was defined in (3.5).

Proof. Apply Lemma 3.8 to the polynomial $f(t) = 1 + \lambda^1(x)t + \cdots + \lambda^n(x)t^n$, and use the definition (3.1) of ψ^n. □

In addition to the closed formula of Theorem 3.9, there is also an inductive type formula relating ψ^i and λ^i for $i \leq k$. This will be used in the next section when we discuss Wilkerson's Theorem.

Theorem 3.10 (Newton Formula). *In any λ-ring R, the relation*

$$\psi^k(x) - \lambda^1(x)\psi^{k-1}(x) + \cdots + (-1)^{k-1}\lambda^{k-1}(x)\psi^1(x) = (-1)^{k+1}k\lambda^k(x)$$

holds for $x \in R$ and $k \geq 1$.

Proof. From the definition (3.1) of Adams operations, we have

$$\lambda_t(x)\psi_{-t}(x) + t\frac{d}{dt}(\lambda_t(x)) = 0.$$

Expanding each of the power series, we obtain

$$\left(\sum_{n\geq 0}\lambda^n(x)t^n\right)\left(\sum_{m\geq 1}(-1)^m\psi^m(x)t^m\right) + \sum_{k\geq 1}k\lambda^k(x)t^k = 0.$$

Now we collect the coefficients of t^k, which gives

$$\sum_{k\geq 1}\left(\sum_{i=0}^{k-1}(-1)^{k-i}\lambda^i(x)\psi^{k-i}(x) + k\lambda^k(x)\right)t^k = 0.$$

So we have

$$\sum_{i=0}^{k-1}(-1)^i\lambda^i(x)\psi^{k-i}(x) = (-1)^{k+1}k\lambda^k(x),$$

which is the required formula. □

Example 3.11. Here are the Newton Formulas for some small values of k:

$$\begin{aligned}
\psi^1(x) &= \lambda^1(x) = x,\\
\psi^2(x) - \lambda^1(x)\psi^1(x) &= -2\lambda^2(x),\\
\psi^3(x) - \lambda^1(x)\psi^2(x) + \lambda^2(x)\psi^1(x) &= 3\lambda^3(x),\\
\psi^4(x) - \lambda^1(x)\psi^3(x) + \lambda^2(x)\psi^2(x) - \lambda^3(x)\psi^1(x) &= -4\lambda^4(x).
\end{aligned}$$

The Newton Formulas can be inverted to write λ^n in terms of $\psi^1, \ldots, \psi^n$, provide that R is free of $\mathbf{Z}$-torsion.

Definition 3.12. A ring R is said to be $\mathbf{Z}$-*torsion-free* if, whenever $na = 0$ for some element $a \in R$ and integer $n \neq 0$, then $a = 0$.

Example 3.13. Here are some examples of $\mathbf{Z}$-torsion-free rings.

(1) Polynomial rings and power series rings over $\mathbf{Z}$ are $\mathbf{Z}$-torsion-free.
(2) Any subring of the rationals $\mathbf{Q}$ is $\mathbf{Z}$-torsion-free.
(3) Any $\mathbf{Q}$-algebra is $\mathbf{Z}$-torsion-free.

Example 3.14. The Newton Formulas (Theorem 3.10) can be used inductively to solve for λ^n in terms of $\psi^1, \ldots, \psi^n$ in a **Z**-torsion-free λ-ring R. For example, we have

$$\lambda^1(x) = \psi^1(x) = x,$$

and

$$\lambda^2(x) = \frac{x^2 - \psi^2(x)}{2}.$$

Putting this into the Newton formula when $k = 3$, we obtain

$$\begin{aligned} \lambda^3(x) &= \frac{1}{3}\left(\psi^3(x) - x\psi^2(x) + \frac{x}{2}(x^2 - \psi^2(x))\right) \\ &= \frac{x^3}{6} - \frac{x\psi^2(x)}{2} + \frac{\psi^3(x)}{3}. \end{aligned}$$

Likewise, putting the expressions for $\lambda^2(x)$ and $\lambda^3(x)$ into the Newton formula when $k = 4$, we obtain

$$\lambda^4(x) = \frac{x^4}{24} + \frac{x\psi^3(x)}{12} - \frac{x^2\psi^2(x)}{4} - \frac{\left(\psi^2(x)\right)^2}{8} - \frac{\psi^4(x)}{4}.$$

Moreover, if we set $\deg(\psi^j(x)) = j$, then $\lambda^k(x)$ is a homogeneous polynomial with rational coefficients of degree k in $\psi^1(x), \ldots, \psi^k(x)$.

The above example illustrates an important property of Adams operations, which we record as follows.

Theorem 3.15. *In a* **Z***-torsion-free λ-ring R, by inductively inverting the Newton Formulas, the λ-operation λ^k $(k \geq 1)$ can be written as a homogeneous polynomial with rational coefficients of degree k in $\psi^1, \ldots, \psi^k$, where ψ^j is given degree j. In particular, when R is* **Z***-torsion-free, the λ-ring structure on R is completely determined by its Adams operations.*

Proof. The proof is by induction on $k \geq 1$. Example 3.14 shows that the Theorem is true for $1 \leq k \leq 4$. Pick an element $x \in R$. For higher values of k, if $i \leq k-1$, then the induction hypothesis says that $\lambda^i(x)$ is a homogenous polynomial with rational coefficients of degree i in $\psi^1(x), \ldots, \psi^i(x)$. Therefore, the product $\lambda^i(x)\psi^{k-i}(x)$ is a homogenous polynomial in $\psi^1(x), \ldots, \psi^k(x)$ with rational coefficients of degree k. By the Newton Formula 3.10 it follows that the same is true for

$$\lambda^k(x) = \frac{(-1)^{k+1}}{k}\sum_{i=0}^{k-1}(-1)^i\lambda^i(x)\psi^{k-i}(x).$$

The **Z**-torsion-free assumption on R is used to make sure that we can unambiguously divide $k\lambda^k(x)$ by k. □

Another application of the Newton Formula is that, in the absence of **Z**-torsion, a ring homomorphism that commutes with all the Adams operations is a λ-homomorphism.

Corollary 3.16. *Let $f\colon R \to S$ be a ring homomorphism between two λ-rings in which S is $\mathbf{Z}$-torsion-free. Suppose that $f\psi^n = \psi^n f$ for all $n \geq 1$. Then f is a λ-homomorphism.*

Proof. It is clear that f commutes with λ^1, which is the identity map. Suppose by induction that $f\lambda^k = \lambda^k f$ for $k < n$. Let x be an element in R. By the Newton Formula (Theorem 3.10) and the induction hypothesis, we have

$$\begin{aligned}
(-1)^{n+1} n f\left(\lambda^n(x)\right) &= f\left((-1)^{n+1} n \lambda^n(x)\right) \\
&= f\left(\sum_{i=0}^{n-1} (-1)^i \lambda^i(x) \psi^{n-i}(x)\right) \\
&= \sum_{i=0}^{n-1} (-1)^i \lambda^i\left(f(x)\right) \psi^{n-i}\left(f(x)\right) \\
&= (-1)^{n+1} n \lambda^n\left(f(x)\right).
\end{aligned}$$

Since S is $\mathbf{Z}$-torsion-free by assumption, it follows that $f\left(\lambda^n(x)\right) = \lambda^n\left(f(x)\right)$. This finishes the induction and the proof of the Corollary. □

3.2 γ-Operations

In this section we discuss basic properties of γ-operations in a λ-ring. These operations were first introduced by Berthelot [Berthelot (1971)] and Grothendieck [Grothendieck (1971)]. Further relations between γ-operations and Adams operations are discussed in the next section.

The materials in this and the next sections mostly follow [Atiyah and Tall (1969)]. The reader is referred to [Husemoller (1994); Osse (1997)] for applications of γ-operations in the representation theory of compact Lie groups. Some of the uses of γ-operations in topological K-theory are discussed in [Atiyah (1989)]. Discussion of γ-operations in the context of the Grothendieck-Riemann-Roch Theorem can be found in [Berthelot *et al.* (1971); Fulton and Lang (1985)].

Definition 3.17. Let R be a λ-ring. Define operations $\gamma^n\colon R \to R$ for $n \geq 0$ by setting

$$\gamma_t(x) = \lambda_{t/(1-t)}(x) \tag{3.6}$$

for $x \in R$, where, as usual, $\gamma_t(x) = \sum_{n\geq 0} \gamma^n(x) t^n$. We call γ^n the nth *γ-operation* on R. If $\gamma_t(x)$ is a polynomial of degree n, then we say that x has *γ-dimension* n.

Here are some basic properties of the γ-operations.

Proposition 3.18. *The following statements hold for elements x and y in a λ-ring R.*

(1) $\lambda_t(x) = \gamma_{t/(1+t)}(x)$.
(2) $\gamma^0(x) = 1$.
(3) $\gamma^1(x) = x$.
(4) If m is an integer, then $\gamma_t(m) = (1-t)^{-m}$. It follows that for $n > m \geq 0$, one has $\gamma^n(-m) = 0$.
(5) One has $\gamma_t(x+y) = \gamma_t(x)\gamma_t(y)$, or, equivalently,

$$\gamma_n(x+y) = \sum_{i=0}^{n} \gamma^i(x)\gamma^{n-i}(y). \tag{3.7}$$

In particular, we have $\gamma_t(-x) = \gamma_t(x)^{-1}$.

Proof. Statement (1) follows from the fact that, if $u = t/(1+t)$, then $u/(1-u) = t$. Statements (2) and (3) both follow from the expansion:

$$\begin{aligned}\lambda_{t/(1-t)}(x) &= \sum_{n\geq 0} \lambda^n(x)t^n(1-t)^{-n} \\ &= 1 + \lambda^1(x)(t + t^2 + \cdots) + \lambda^2(x)(t^2 + \cdots) + \cdots \\ &\equiv 1 + xt \pmod{t^2}.\end{aligned}$$

For statement (4), recall that $\lambda_t(m) = (1+t)^m$ for an integer m (Example 1.15). Replacing t by $t/(1-t)$, we obtain

$$\gamma_t(m) = \left(1 + \frac{t}{1-t}\right)^m = (1-t)^{-m}.$$

For statement (5), we have

$$\begin{aligned}\gamma_t(x+y) &= \lambda_{t/(1-t)}(x+y) \\ &= \sum_{n\geq 0}\Bigl(\sum_{i+j=n} \lambda^i(x)\lambda^j(y)\Bigr)t^n(1-t)^{-n} \\ &= \Bigl(\sum_{i\geq 0}\lambda^i(x)t^i(1-t)^{-i}\Bigr)\Bigl(\sum_{j\geq 0}\lambda^j(y)t^j(1-t)^{-j}\Bigr) \\ &= \gamma_t(x)\gamma_t(y)\end{aligned}$$

Finally, using statement (4), we have

$$1 = \gamma_t(x-x) = \gamma_t(x)\gamma_t(-x),$$

as desired. □

The following result shows that the operation γ^n is the composition of an integer translation and λ^n.

Proposition 3.19. *In a λ-ring R, we have $\gamma^n(x) = \lambda^n(x + n - 1)$ for $x \in R$ and $n \geq 0$.*

Proof. First we compute as follows:

$$\begin{aligned}\gamma_t(x) &= \gamma_t((x+n-1)-(n-1)) \\ &= \gamma_t(x+n-1)\gamma_t(1)^{-(n-1)} \\ &= \lambda_{t/(1-t)}(x+n-1)\cdot(1-t)^{n-1} \\ &= \sum_{i\geq 0}\lambda^i(x+n-1)t^i(1-t)^{n-1-i}.\end{aligned} \tag{3.8}$$

In this sum, for $i < n$, the highest power of t in

$$t^i(1-t)^{n-1-i} = t^i(1+\cdots+(-1)^{n-1-i}t^{n-1-i})$$

is t^{n-1}. So it does not involve t^n. Likewise, if $i > n$, then

$$\begin{aligned}t^i(1-t)^{n-1-i} &= t^i\left(1-(n-1-i)t+\cdots\right) \\ &= t^i-(n-1-i)t^{i+1}+\cdots,\end{aligned}$$

which also does not involve t^n. When $i = n$, we have

$$t^n(1-t)^{-1} = t^n + t^{n+1} + \cdots.$$

Therefore, on the right-hand side of (3.8), the only term that involves t^n is

$$\lambda^n(x+n-1)t^n(1-t)^{-1},$$

from which the result follows. □

The previous Proposition implies that γ^n can be expressed as a **Z**-linear combination of the λ-operations, and vice versa.

Corollary 3.20. *The following statements hold for elements x and y in a λ-ring R and $n \geq 1$.*

*(1) The γ-operation γ^n is a **Z**-linear combination of $\lambda^1, \ldots, \lambda^n$:*

$$\gamma^n(x) = \sum_{i=0}^{n-1}\binom{n-1}{i}\lambda^{n-i}(x). \tag{3.9}$$

In particular, γ^n is a natural operation on λ-rings (Definition 1.40).

*(2) The λ-operation λ^n is a **Z**-linear combination of $\gamma^1, \ldots, \gamma^n$:*

$$\lambda^n(x) = \sum_{i=0}^{n-1}(-1)^i\binom{n-1}{i}\gamma^{n-i}(x). \tag{3.10}$$

Proof. For statement (1), by Proposition 3.19 and axiom (4) in Definition 1.10, we have

$$\begin{aligned}\gamma^n(x) &= \lambda^n(x+(n-1)) \\ &= \lambda^n(x)+\lambda^1(n-1)\lambda^{n-1}(x)+\cdots+\lambda^{n-1}(n-1)\lambda^1(x),\end{aligned}$$

which is the required formula (3.9).

For statement (2), by Proposition 3.19 and statement (5) in Proposition 3.18, we have

$$\begin{aligned}\lambda^n(x) &= \gamma^n(x + (1-n)) \\ &= \sum_{i=0}^{n-1} \gamma^i(1-n)\gamma^{n-i}(x).\end{aligned}$$

Now observe that $\gamma^i(1-n)$, by statement (4) in Proposition 3.18, is the coefficient of t^i in the polynomial $\gamma_t(1-n) = (1-t)^{n-1}$, which is $(-1)^i\binom{n-1}{i}$, as desired. □

Example 3.21. The first few cases of the formula (3.9) are:

$$\begin{aligned}\gamma^1(x) &= \lambda^1(x), \\ \gamma^2(x) &= \lambda^2(x) + \lambda^1(x), \\ \gamma^3(x) &= \lambda^3(x) + 2\lambda^2(x) + \lambda^1(x), \\ \gamma^4(x) &= \lambda^4(x) + 3\lambda^3(x) + 3\lambda^2(x) + \lambda^1(x), \\ \gamma^5(x) &= \lambda^5(x) + 4\lambda^4(x) + 6\lambda^3(x) + 4\lambda^2(x) + \lambda^1(x).\end{aligned}$$

Likewise, the first few cases of the formula (3.10) are:

$$\begin{aligned}\lambda^1(x) &= \gamma^1(x), \\ \lambda^2(x) &= \gamma^2(x) - \gamma^1(x), \\ \lambda^3(x) &= \gamma^3(x) - 2\gamma^2(x) + \gamma^1(x), \\ \lambda^4(x) &= \gamma^4(x) - 3\gamma^3(x) + 3\gamma^2(x) - \gamma^1(x), \\ \lambda^5(x) &= \gamma^5(x) - 4\gamma^4(x) + 6\gamma^3(x) - 4\gamma^2(x) + \gamma^1(x).\end{aligned}$$

Corollary 3.22. *Let I be a λ-ideal in a λ-ring R. Then I is closed under the γ-operations, in the sense that $\gamma^n(I) \subseteq I$ for $n \geq 1$.*

Proof. If $x \in I$ and $n \geq 1$, then (3.9) expresses $\gamma^n(x)$ as a finite sum of elements of the form $\lambda^j(x) \in I$ for $1 \leq j \leq n$. Thus, $\gamma^n(x)$ lies in I as well. □

The dimension of an element in a λ-ring is related to its γ-dimension as follows.

Corollary 3.23. *Let x be an element in a λ-ring. Then x has dimension at most 1 if and only if $(x-1)$ has γ-dimension at most 1.*

Proof. First suppose that x has dimension at most 1. By Proposition 3.19 we have that for $n \geq 2$,

$$\begin{aligned}\gamma^n(x-1) &= \lambda^n(x + (n-2)) \\ &= \sum_{i=0}^{n} \lambda^i(n-2)\lambda^{n-i}(x).\end{aligned}$$

For $i \leq n-2$, we know that $\lambda^{n-i}(x) = 0$. For $i > n-2$, we have $\lambda^i(n-2) = \binom{n-2}{i} = 0$. Therefore, $(x-1)$ has γ-dimension at most 1.

Conversely, suppose that x has γ-dimension at most 1. By Proposition 3.19 again and Proposition 3.18, we have that for $n \geq 2$,

$$\begin{aligned}\lambda^n(x+1) &= \gamma^n(x+(2-n)) \\ &= \sum_{i=0}^{n} \gamma^i(2-n)\gamma^{n-i}(x) \\ &= \sum_{i=0}^{n} (-1)^i \binom{n-2}{i} \gamma^{n-i}(x).\end{aligned}$$

We know that $\gamma^{n-i}(x) = 0$ for $i \leq n-2$. For $i > n-2$, the binomial coefficient $\binom{n-2}{i} = 0$. Therefore, $(x+1)$ has dimension at most 1. □

Using Corollary 3.20 we can derive the analogs for γ-operations of axioms (5) and (6) in Definition 1.10 of a λ-ring.

Corollary 3.24. *There exist universal polynomials G_n and $G_{n,m}$ $(n, m \geq 1)$ with integer coefficients and zero constant terms such that*

$$\begin{aligned}\gamma^n(xy) &= G_n(\gamma^1(x), \ldots, \gamma^n(x); \gamma^1(y), \ldots, \gamma^n(y)), \\ \gamma^n(\gamma^m(x)) &= G_{n,m}(\gamma^1(x), \ldots, \gamma^{nm}(x))\end{aligned}$$

for elements x and y in any λ-ring R.

Proof. Using (3.9) we write

$$\begin{aligned}\gamma^n(xy) &= \sum_{i=0}^{n-1} \binom{n-1}{i} \lambda^{n-i}(xy) \\ &= \sum_{i=0}^{n-1} \binom{n-1}{i} P_{n-i}(\lambda^1(x), \ldots, \lambda^{n-i}(x); \lambda^1(y), \ldots, \lambda^{n-i}(y)).\end{aligned}$$

Here P_{n-i} is the universal polynomial in $2(n-i)$ variables with integer coefficients and zero constant term that appeared in Definition 1.10 and was first defined in Example 1.9. The above expression involves only $\lambda^1(x), \ldots, \lambda^n(x)$ and $\lambda^1(y), \ldots, \lambda^n(y)$. Now using 3.10 we write each $\lambda^j(x)$ appearing in the above sum as a $\mathbf{Z}$-linear combination of $\gamma^1(x), \ldots, \gamma^j(x)$. The same goes for $\lambda^j(y)$. Therefore, we have written $\gamma^n(xy)$ as a polynomial with integer coefficients and zero constant term in $\gamma^1(x), \ldots, \gamma^n(x)$ and $\gamma^1(y), \ldots, \gamma^n(y)$.

For the second equality, we use (3.9) twice to get

$$\begin{aligned}\gamma^n(\gamma^m(x)) &= \sum_{i=0}^{n-1} \binom{n-1}{i} \lambda^{n-i}(\gamma^m(x)) \\ &= \sum_{i=0}^{n-1} \binom{n-1}{i} \lambda^{n-i}\left(\sum_{j=0}^{m-1} \binom{m-1}{j} \lambda^{m-j}(x)\right).\end{aligned}$$

Using axiom (4) in Definition 1.10 of a λ-ring repeatedly, each summand

$$\lambda^{n-i}\Big(\sum_{j=0}^{m-1}\binom{m-1}{j}\lambda^{m-j}(x)\Big)$$

becomes a polynomial with integer coefficients and zero constant term in $\lambda^p(\lambda^q(x))$ for $p \leq n$ and $q \leq m$. But since

$$\lambda^p(\lambda^q(x)) = P_{p,q}(\lambda^1(x), \ldots, \lambda^{pq}(x))$$

by axiom (6) in Definition 1.10, using (3.10) we can write $\lambda^p(\lambda^q(x))$ as a polynomial with integer coefficients and zero constant term in $\gamma^1(x), \ldots, \gamma^{pq}(x)$. In other words, we have written $\gamma^n(\gamma^m(x))$ as a polynomial with integer coefficients and zero constant term in $\gamma^1(x), \ldots, \gamma^{nm}(x)$. □

Example 3.25. We can find the polynomial G_2 as follows:

$$\begin{aligned}
\gamma^2(xy) &= \lambda^2(xy) + \lambda^1(xy) \\
&= xy + P_2(x, \lambda^2(x); y, \lambda^2(y)) \\
&= xy + x^2\lambda^2(y) + y^2\lambda^2(x) - 2\lambda^2(x)\lambda^2(y) \\
&= xy + x^2(\gamma^2(y) - y) + y^2(\gamma^2(x) - x) - 2(\gamma^2(x) - x)(\gamma^2(y) - y) \\
&= -xy - x^2y - xy^2 + x(x+2)\gamma^2(y) + y(y+2)\gamma^2(x) - 2\gamma^2(x)\gamma^2(y).
\end{aligned}$$

In other words, we have

$$\begin{aligned}
G_2(x_1, x_2; y_1, y_2) = -x_1y_1 - x_1^2y_1 - x_1y_1^2 + x_1(x_1+2)y_2 \\
+ y_1(y_1+2)x_2 - 2x_2y_2.
\end{aligned}$$

In general, for $n \geq 2$, we use Corollary 3.20 to obtain

$$G_n(x_1, \ldots, x_n; y_1, \ldots, y_n) = \sum_{i=0}^{n-1}\binom{n-1}{i}P_{n-i}(a_1, \ldots, a_{n-i}; b_1, \ldots, b_{n-i}),$$

where

$$a_j = \sum_{k=0}^{j-1}(-1)^k\binom{j-1}{k}x_{j-k} \quad \text{and} \quad b_j = \sum_{k=0}^{j-1}(-1)^k\binom{j-1}{k}y_{j-k}.$$

3.3 γ-Filtration

In an augmented λ-ring, such as the K-theory of a pointed space and the representation ring of a group (Examples 1.31 and 1.32), the γ-operations give rise to the γ-filtration, which is preserved by Adams operations. There is an analog of The Verification Principle for γ-rings (Theorem 3.40). Moreover, the Adams operation ψ^k acts as multiplication by k^n on the nth quotient of the γ-filtration (Theorem 3.41).

Let R be an augmented λ-ring (Definition 1.30) with augmentation $\varepsilon\colon R \to \mathbf{Z}$ and augmentation ideal $I = \ker \varepsilon$. This means that I is a λ-ideal in R, and there is a decomposition $R = \mathbf{Z} \oplus I$ as an abelian group (Proposition 1.33).

Definition 3.26. For $n \geq 0$ let I_n denote the additive subgroup of R generated by products $\gamma^{n_1}(a_1) \cdots \gamma^{n_r}(a_r)$, where each $a_i \in I = \ker \varepsilon$ and $n_1 + \cdots + n_r \geq n$. We call I_n the nth *γ-filtration* of R.

Proposition 3.27. *The following statements hold in an augmented λ-ring R:*

(1) One has

$$I_0 \supseteq I_1 \supseteq I_2 \supseteq \cdots. \tag{3.11}$$

(2) One has $I_0 = R$, and $I_1 = I$.
(3) For $m, n \geq 0$, one has $I_m \cdot I_n \subseteq I_{m+n}$.
(4) Each I_n is a λ-ideal in R.

Proof. Statements (1) and (3) are obvious from the definition.

For statement (2), since γ^1 is the identity map (Proposition 3.18), it follows that I_1 contains I. Conversely, we know that each $\gamma^i(a) \in I$ for $a \in I$ and $i \geq 1$ (Corollary 3.22). It follows that I_1 is contained in I, so I_1 is actually equal to I. Moreover, since $\gamma^0(x) = 1$ for any $x \in R$, I_0 contains both $I_1 = I$ and the integers $\mathbf{Z}$ in R. Since $R = \mathbf{Z} \oplus I$, it follows that I_0 is R itself.

For statement (4), first observe that I_n is an ideal in R, since every element in R can be written as a sum of an integer and an element in $I = I_1$. To show that I_m is a λ-ideal, it suffices to show that

$$\lambda^n(\gamma^m(x)) \in I_m \tag{3.12}$$

for $x \in I$ and $n, m \geq 1$. Using Proposition 3.19 we have

$$\begin{aligned}
\lambda^n(\gamma^m(x)) &= \lambda^n(\lambda^m(x + m - 1)) \\
&= P_{n,m}(\lambda^1(x + m - 1), \ldots, \lambda^{nm}(x + m - 1)).
\end{aligned}$$

By Lemma 1.45 every non-zero term in the polynomial $P_{n,m}(s_1, \ldots, s_{nm})$ contains a factor of s_i for some $i \geq m$, where the s_k are the elementary symmetric functions of the variables $x_1, \ldots, x_{nm}$. Therefore, to prove (3.12), it suffices to show that

$$\lambda^i(x + m - 1) \in I_m \tag{3.13}$$

for $i \geq m$. Let $s = i - m \geq 0$. Then by statement (5) of Proposition 3.18 and Proposition 3.19, we have

$$\begin{aligned}
\lambda^i(x + m - 1) &= \lambda^{m+s}(x + m - 1) \\
&= \gamma^{m+s}((x + m - 1) - (m + s) + 1) \\
&= \gamma^{m+s}(x - s) \\
&= \sum_{k=0}^{m+s} \gamma^{m+s-k}(x)\gamma^k(-s).
\end{aligned}$$

By statement (4) of Proposition 3.18, we know that $\gamma^k(-s) = 0$ for $k > s \geq 0$. Therefore, we have

$$\lambda^i(x+m-1) = \sum_{k=0}^{s} \gamma^{m+s-k}(x)\gamma^k(-s).$$

Now for $0 \leq k \leq s$, $\gamma^k(-s)$ is an integer, and $\gamma^{m+s-k}(x) \in I_{m+s-k} \subseteq I_m$. This proves (3.13) and that I_n is a λ-ideal. □

Definition 3.28. For an augmented λ-ring R with augmentation ideal I, the filtration

$$R \supseteq I \supseteq I_2 \supseteq I_3 \supseteq \cdots$$

in Proposition 3.27 is called the *γ-filtration of R* (or of I).

In other words, Proposition 3.27 says that the γ-filtration is a decreasing and multiplicative filtration of R by λ-ideals. The γ-filtration actually makes an augmented λ-ring into a *filtered* λ-ring (Example 6.6), which is the subject of Chapters 6 and 7.

Example 3.29. Recall from Example 1.31 that the topological K-theory $K(X)$ of a pointed space X is an augmented λ-ring. Therefore, it has an associated γ-filtration. Likewise, the representation ring $R(G)$ of a group G is an augmented λ-ring (Example 1.32), so it also has an associated γ-filtration.

Definition 3.30. By a *γ-ring*, we mean the augmentation ideal I of an augmented λ-ring R. We consider I as a non-unital ring equipped with the γ-operations $\gamma^n \colon I \to I$ $(n \geq 1)$ of R. A *γ-homomorphism* of γ-rings is a homomorphism $f \colon I \to I'$ of non-unital rings that commutes with the γ-operations, in the sense that the square

$$\begin{array}{ccc} I & \xrightarrow{f} & I' \\ {\scriptstyle \gamma_n}\big\downarrow & & \big\downarrow{\scriptstyle \gamma_n} \\ I & \xrightarrow{f} & I' \end{array}$$

commutes for $n \geq 1$.

Proposition 3.31. *The following statements about the γ-filtration hold.*

(1) Any γ-homomorphism preserves the γ-filtration. In other words, if $f \colon I \to I'$ is a γ-homomorphism, then $f(I_n) \subseteq I'_n$ for $n \geq 1$.
(2) Each I_n is closed under γ^i, in the sense that $\gamma^i(I_n) \subseteq I_n$ for $i \geq 1$.

Proof. Statement (1) is true because a γ-homomorphism is a non-unital ring homomorphism that also commutes with the γ-operations. For statement (2), first recall that γ^i is a $\mathbf{Z}$-linear combination of $\lambda^1, \ldots, \lambda^i$ (3.9). The result follows, since I_n is a λ-ideal by statement (4) in Proposition 3.27. □

Example 3.32. Every λ-homomorphism $\varphi\colon R \to S$ of augmented λ-rings is a γ-homomorphism, and, therefore, it preserves the γ-filtration. In fact, since each γ-operation γ^n $(n \geq 1)$ is a $\mathbf{Z}$-linear combination of $\lambda^1, \ldots, \lambda^n$ (3.9), we know that φ commutes with γ^n and $\varphi(I_n) \subseteq I_n$. It follows that φ restricts to a γ-homomorphism $\varphi\colon I_R \to I_S$ of the augmentation ideals I_R and I_S. In particular, every Adams operation ψ^k on an augmented λ-ring R restricts to a γ-homomorphism on the augmentation ideals, and it preserves the γ-filtration.

3.3.1 *Verification principle for γ-rings*

Now we discuss the analog of The Verification Principle (Theorem 1.43) for γ-rings. This is needed in order to describe the nice action of ψ^k on the γ-filtration.

Definition 3.33. By a *natural operation on γ-rings*, we mean a rule μ that assigns to each γ-ring I a function $\mu\colon I \to I$ such that, if $\varphi\colon I_R \to I_S$ is a γ-homomorphism, then it is required that the diagram

$$\begin{array}{ccc} I_R & \xrightarrow{\varphi} & I_S \\ {\scriptstyle\mu}\downarrow & & \downarrow{\scriptstyle\mu} \\ I_R & \xrightarrow{\varphi} & I_S \end{array}$$

be commutative.

Example 3.34. Every γ-operation γ^n $(n \geq 1)$ is a natural operation on γ-rings by definition. By (3.10) so is λ^n for $n \geq 1$. Moreover, since the Adams operation ψ^k is a homogeneous polynomial with integer coefficients and zero constant term in $\lambda^1, \ldots, \lambda^k$ (Theorem 3.9), it is also a natural operation on γ-rings.

In fact, any polynomial with integer coefficients and zero constant term of natural operations on γ-rings is again a natural operation on γ-rings. Therefore, natural operations on γ-rings form a non-unital ring, which is denoted by Op^γ. There is a homomorphism

$$\alpha^\gamma\colon \mathbf{Z}[\gamma^1, \gamma^2, \ldots]^+ \to Op^\gamma \tag{3.14}$$

of non-unital rings. Here $\mathbf{Z}[\gamma^1, \gamma^2, \ldots]^+$ is the non-unital ring of polynomials with integer coefficients and zero constant term in the γ-operations γ^n $(n \geq 1)$.

The following discussion about free γ-rings and natural operations on γ-rings is the obvious analog of §1.3 and §1.4 for γ-rings. We will omit the proofs, since they are essentially identical to their λ-counterparts.

Let $x_1, x_2, \ldots$ be an infinite sequence of variables. Consider the non-unital ring

$$\Omega_r^\gamma = \mathbf{Z}[x_1, \ldots, x_r]^+$$

of polynomials with integer coefficients and zero constant term. It is the augmentation ideal of the full polynomial ring $\mathbf{Z}[x_1, \ldots, x_r]$ and is a γ-ring with

$\gamma_t(x_i) = 1 + x_i t$ for all i, i.e., each x_i has γ-dimension 1. There is a γ-homomorphism $p_r\colon \Omega_r^\gamma \to \Omega_{r-1}^\gamma$ defined by

$$p_r(x_i) = \begin{cases} x_i & \text{if } 1 \le i \le r-1, \\ 0 & \text{if } i = r. \end{cases}$$

The inverse system

$$\Omega_1^\gamma \xleftarrow{p_2} \Omega_2^\gamma \xleftarrow{p_3} \Omega_3^\gamma \xleftarrow{p_4} \cdots$$

has an inverse limit $\Omega^\gamma = \varprojlim_r \Omega_r^\gamma$, which is also a γ-ring. An element in Ω^γ is a power series over $\mathbf{Z}$ with 0 constant term in the variables x_i such that, setting $x_{r+1} = x_{r+2} = \cdots = 0$, it becomes a polynomial with 0 constant term in $x_1, \ldots, x_r$ $(r \ge 1)$.

Let $s_n(x_1, \ldots, x_r)$ be the nth elementary symmetric function on the variables $x_1, \ldots, x_r$. Then

$$s_n = \varprojlim_r s_n(x_1, \ldots, x_r) = \sum_{i_1 < \cdots < i_n} x_{i_1} \cdots x_{i_n}$$

is an element in Ω^γ and is thought of as the nth elementary symmetric function on the infinite sequence of variables $x_1, x_2, \ldots$.

The following result is the analog of Proposition 1.36, whose proof can be adapted by replacing Ω with Ω^γ and λ^n by γ^n.

Proposition 3.35. *In the γ-ring Ω^γ, the elements s_n $(n \ge 1)$ are algebraically independent. Moreover, one has $\gamma^n(s_1) = s_n$ for $n \ge 1$.*

Definition 3.36. Let U^γ be the smallest γ-ring in Ω^γ that contains s_1. The γ-ring U^γ is called the *free γ-ring on one generator.*

Clearly U^γ is the γ-ring analog of the free λ-ring U on one generator (Definition 1.37).

The following result is the analog of Proposition 1.38, whose proof can be adapted by replacing U with U^γ and λ^n with γ^n. One also needs to use Corollary 3.24 to see that the composition $\gamma^n(\gamma^m(x))$ is a polynomial with integer coefficients and zero constant term in $\gamma^i(x)$ for $1 \le i \le nm$.

Proposition 3.37. *The free γ-ring on one generator is the non-unital ring*

$$U^\gamma = \mathbf{Z}[s_1, s_2, s_3, \ldots]^+$$

of integer polynomials with 0 constant term on the algebraically independent elements s_i $(i \ge 1)$.

The free γ-ring has the following universal property, which is the analog of Theorem 1.39.

Theorem 3.38. *Let I be a γ-ring, and let r be an element in I. Then there exists a unique γ-homomorphism $u_r^\gamma\colon U^\gamma \to I$ such that $\gamma^n(r) = u_r^\gamma(s_n)$ for $n \ge 1$.*

This universal property of the free γ-ring is used to prove the following result, which is the analog of Theorem 1.42. Again, the proof is adapted from that of Theorem 1.42 by replacing U with U^γ and λ^n with γ^n. Moreover, a natural operation on γ-rings is completely determined by its action on the element $s_1 \in U^\gamma$.

Theorem 3.39. *The homomorphism* (3.14) $\alpha^\gamma \colon \mathbf{Z}[\gamma^1, \gamma^2, \ldots]^+ \to Op^\gamma$ *of non-unital rings is an isomorphism.*

In other words, every natural operation on γ-rings is uniquely a polynomial with integer coefficients and zero constant term in the γ-operations γ^i $(i \geq 1)$.

As in the case of natural operations on λ-rings, Theorem 3.39 leads to the following analog of The Verification Principle (Theorem 1.43). It says that a natural operation μ on γ-rings is uniquely determined by the elements $\mu(x_1 + \cdots + x_r)$, where $r \geq 1$ and each x_i has γ-dimension one.

Theorem 3.40 (The Verification Principle for γ-rings). *Every natural operation μ on γ-rings is uniquely a polynomial with integer coefficients and* 0 *constant term in the γ-operations γ^i $(i \geq 1)$. Moreover, one has $\mu = f(\gamma^1, \ldots, \gamma^n)$ if and only if this equality holds when applied to finite sums of elements of γ-dimension one.*

Natural operations on γ-rings with multiple inputs are defined similarly. Indeed, a natural operation on γ-rings with $r \geq 1$ inputs is a rule μ that assigns to each γ-ring I a function

$$\mu \colon I^r = \underbrace{I \times \cdots \times I}_{r \text{ times}} \to I$$

such that, if $\varphi \colon I_R \to I_S$ is a γ-homomorphism, then the square

$$\begin{array}{ccc} I_R^r & \xrightarrow{\varphi^r} & I_S^r \\ \mu \downarrow & & \downarrow \mu \\ I_R & \xrightarrow{\varphi} & I_S \end{array}$$

commutes.

Starting with r infinite sequences $\{x_i^1\}, \ldots, \{x_i^r\}$ of variables, there is a corresponding free γ-ring

$$U_r^\gamma = \mathbf{Z}[s_1^1, s_2^1, \ldots; \ldots; s_1^r, s_2^r, \ldots]^+ \tag{3.15}$$

on r generators. Here s_j^k is the jth elementary symmetric function on the variables $\{x_i^k\}$. It has the right universal property and can be used to prove the corresponding Verification Principle for natural operations on γ-rings with r inputs. We leave it as an exercise for the reader to formulate and prove these results.

3.3.2 *Adams operations acting on the γ-filtration*

Now we describe how Adams operations act on the γ-filtration.

Theorem 3.41. *Let R be an augmented λ-ring with augmentation ideal I. For $k, n \geq 1$, we have*

$$\psi^k(x) - k^n x \in I_{n+1} \tag{3.16}$$

for $x \in I_n$, the nth γ-filtration of R (Definition 3.26).

Proof. Since the Adams operation ψ^k is multiplicative (Theorem 3.6), it suffices to show that

$$\psi^k(\gamma^m(x)) - k^m(\gamma^m(x)) \in I_{m+1} \tag{3.17}$$

for $x \in I$. Since $\psi^k\gamma^m - k^m\gamma^m$ is a natural operation on γ-rings, by The Verification Principle for γ-rings (Theorem 3.40) it suffices to prove (3.17) when x is a finite sum of elements of γ-dimension one.

If $x_i \in I$ has γ-dimension one, then $(1+x_i)$ has dimension at most 1 by Corollary 3.23. It follows from Proposition 3.5 that

$$\psi^k(1 + x_i) = (1 + x_i)^k,$$

which is equivalent to

$$\psi^k(x_i) = (1 + x_i)^k - 1$$

because ψ^k is a ring homomorphism. Now let $x_1, \ldots, x_r$ be elements in I of γ-dimension one. By Proposition 3.18 we have

$$\begin{aligned}
&(\psi^k\gamma^m - k^m\gamma^m)(x_1 + \cdots + x_r) \\
&\quad = \psi^k\left(s_m(x_1, \ldots, x_r)\right) - k^m s_m(x_1, \ldots, x_r) \\
&\quad = s_m(\psi^k(x_1), \ldots, \psi^k(x_r)) - k^m s_m(x_1, \ldots, x_r) \\
&\quad = s_m\left((1 + x_1)^k - 1, \ldots, (1 + x_r)^k - 1\right) - k^m s_m(x_1, \ldots, x_r) \\
&\quad = \sum_{1 \leq i_1 < \cdots < i_m \leq r} \left(kx_{i_1} + \binom{k}{2}x_{i_1}^2 + \cdots\right) \cdots \left(kx_{i_m} + \binom{k}{2}x_{i_m}^2 + \cdots\right) \\
&\qquad - \sum_{1 \leq i_1 < \cdots < i_m \leq r} k^m x_{i_1} \cdots x_{i_m}.
\end{aligned}$$

Here $s_m(x_1, \ldots, x_r)$ denotes the mth elementary symmetric function on $x_1, \ldots, x_r$. This shows that

$$(\psi^k\gamma^m - k^m\gamma^m)(x_1 + \cdots + x_r)$$

is an integer polynomial in $x_1, \ldots, x_r$ in which each non-zero monomial is of degree at least $m + 1$ in the x_i. This proves (3.17) when x is a finite sum of elements of γ-dimension one, as desired. □

In other words, the Adams operation ψ^k induces an additive map

$$\psi^k\colon I_n/I_{n+1} \to I_n/I_{n+1}$$

on the nth associated quotient of the γ-filtration of R, and this induced map is equal to multiplication by k^n.

Moreover, on the nth associated quotient of the γ-filtration, the Adams operation ψ^k differs from the λ-operation λ^k only by a factor of $\pm k$, as the following result shows.

Proposition 3.42. *Let R be an augmented λ-ring with augmentation ideal I, and let x be an element of the nth γ-filtration I_n $(n \geq 1)$. Then we have*

$$\psi^k(x) + (-1)^k k\lambda^k(x) \in I_{n+1}. \tag{3.18}$$

Proof. Since $x \in I_n$, we know that both $\psi^i(x)$ and $\lambda^i(x)$ lie in I_n for $i \geq 1$. From the Newton Formula (Theorem 3.10)

$$\psi^k(x) - \lambda^1(x)\psi^{k-1}(x) + \cdots + (-1)^{k-1}\lambda^{k-1}(x)\psi^1(x) = (-1)^{k+1}k\lambda^k(x),$$

we see that

$$\psi^k(x) + (-1)^k k\lambda^k(x) = \lambda^1(x)\psi^{k-1}(x) + \cdots + (-1)^k\lambda^{k-1}(x)\psi^1(x)$$

is an element of I_{2n}, which is a subset of I_{n+1} for $n \geq 1$. □

The previous two results can be combined to obtain information about λ^k.

Corollary 3.43. *Let R be an augmented λ-ring with augmentation ideal I, and let x be an element of the nth γ-filtration I_n $(n \geq 1)$. Then we have $\lambda^k(x) + (-1)^k k^{n-1}x \in I_{n+1}$ for $k \geq 1$.*

Proof. First note that for $a, b \in I_n$, we have

$$\begin{aligned}\lambda^n(a+b) &= \lambda^n(a) + \lambda^n(b) + \sum_{i=1}^{n-1}\lambda^i(a)\lambda^{n-i}(b)\\ &\equiv \lambda^n(a) + \lambda^n(b) \pmod{I_{n+1}}.\end{aligned}$$

The congruence follows from the fact that $\lambda^i(a), \lambda^j(b) \in I_n$ for $i, j \geq 1$, so $\lambda^i(a)\lambda^{n-i}(b) \in I_{2n} \subseteq I_{n+1}$. In other words, λ^n induces an additive map $\lambda^n\colon I_n/I_{n+1} \to I_n/I_{n+1}$. Therefore, to prove the Corollary, it suffices to prove

$$\lambda^k(\gamma^{n_1}(a_1)\cdots\gamma^{n_r}(a_r)) + (-1)^k k^{n-1}\gamma^{n_1}(a_1)\cdots\gamma^{n_r}(a_r) \in I_{n+1} \tag{3.19}$$

when each $a_i \in I$ and $n_1 + \cdots + n_r \geq n$. Using (3.10) and Corollary 3.24, we see that

$$\mu(a_1, \ldots, a_r) = \lambda^k(\gamma^{n_1}(a_1)\cdots\gamma^{n_r}(a_r)) + (-1)^k k^{n-1}\gamma^{n_1}(a_1)\cdots\gamma^{n_r}(a_r)$$

is a natural operation on γ-rings with r inputs. Subtracting (3.16) from (3.18) and multiplying by $(-1)^k$, we see that $k\mu(a_1, \ldots, a_r) \in I_{n+1}$. Since the free γ-ring on r generators U_r^γ (3.15) is $\mathbf{Z}$-torsion-free and μ is determined by its action on U_r^γ, it follows that $\mu(a_1, \ldots, a_r) \in I_{n+1}$, as desired. □

3.4 ψ-Rings

In this section, we discuss ψ-rings, which are rings equipped with Adams-type operations, following [Knutson (1973)]. The main result of this section (Theorem 3.49) states that a $\mathbf{Z}$-torsion-free ψ-ring R gives rise to a λ-ring structure on $R \otimes \mathbf{Q}$.

Definition 3.44. A *ψ-ring* is a ring R equipped with ring endomorphisms $\psi^k \colon R \to R$ for $k \geq 1$ such that $\psi^1 = Id$ and $\psi^m \psi^n = \psi^{mn}$ for all $m, n \geq 1$. A *ψ-homomorphism* (respectively, ψ-isomorphism) $f \colon R \to S$ of ψ-rings is a ring homomorphism (respectively, isomorphism) such that the diagram

$$\begin{array}{ccc} R & \xrightarrow{f} & S \\ {\scriptstyle \psi^n}\downarrow & & \downarrow{\scriptstyle \psi^n} \\ R & \xrightarrow{f} & S \end{array}$$

commutes for each $n \geq 1$.

Example 3.45. Every ring R has the trivial ψ-ring structure in which $\psi^n = Id$ for all n.

Example 3.46. A λ-ring R is a ψ-ring in which the ψ^k are the Adams operations of R (Theorems 3.6 and 3.7). A λ-homomorphism is also a ψ-homomorphism, since Adams operations are natural operations on λ-rings.

There is a non-trivial ψ-ring structure on the Cartesian product ring of any given ring. Let R be a ring. The *Cartesian product ring* is defined as the countable product

$$R^{\mathbf{N}} = \prod_{i=1}^{\infty} R. \tag{3.20}$$

An element in $R^{\mathbf{N}}$ is an infinite sequence $\mathbf{a} = (a_1, a_2, \ldots)$, with each $a_i \in R$. Addition and multiplication are defined entry-wise, so its additive zero and multiplicative identity are

$$(0, 0, 0, \ldots) \quad \text{and} \quad (1, 1, 1, \ldots), \tag{3.21}$$

respectively.

Proposition 3.47. *Let R be a ring. Then the Cartesian product ring $R^{\mathbf{N}}$ is a ψ-ring in which $\psi^k \colon R^{\mathbf{N}} \to R^{\mathbf{N}}$ is given by $\psi^k((a_1, a_2, \ldots)) = (a_k, a_{2k}, a_{3k}, \ldots)$ for $k \geq 1$.*

Proof. It is clear that ψ^1 is the identity map and that ψ^k preserves the additive zero and the multiplicative identity (3.21). The additivity and multiplicativity of ψ^k

follow from the fact that addition and multiplication in $R^{\mathbf{N}}$ are defined entry-wise. Finally, we have

$$\begin{aligned}\psi^m(\psi^n((a_1,a_2,\ldots))) &= \psi^m((a_n,a_{2n},a_{3n},\ldots))\\ &= (a_{mn},a_{2mn},a_{3mn},\ldots)\\ &= \psi^{mn}((a_1,a_2,\ldots)),\end{aligned}$$

as desired. □

Proposition 3.48. *Let R be a $\mathbf{Z}$-torsion-free ψ-ring. Then the following statements hold.*

(1) The tensor product $R\otimes\mathbf{Q}$ is a ψ-ring in which $\psi^k(r\otimes q)=\psi^k(r)\otimes q$ for $r\in R$ and $q\in\mathbf{Q}$.
(2) The map $\Psi\colon R\to R^{\mathbf{N}}$ defined by

$$\Psi(a)=(\psi^1(a),\psi^2(a),\psi^3(a),\ldots) \tag{3.22}$$

for $a\in R$ is an injective ψ-homomorphism.

Proof. Statement (1) is immediate from the definition, since ψ^k on $R\otimes\mathbf{Q}$ is defined as the tensor product $\psi^k\otimes Id$.

For statement (2), first observe that Ψ is a ring homomorphism because each ψ^k is and the ring operations in $R^{\mathbf{N}}$ are defined entry-wise. By Proposition 3.47, $R^{\mathbf{N}}$ is a ψ-ring. To see that Ψ is a ψ-homomorphism, observe that

$$\begin{aligned}\psi^k(\Psi(a)) &= \psi^k(\psi^1(a),\psi^2(a),\psi^3(a),\ldots)\\ &= (\psi^k(a),\psi^{2k}(a),\psi^{3k}(a),\ldots)\\ &= (\psi^1(\psi^k(a)),\psi^2(\psi^k(a)),\psi^3(\psi^k(a)),\ldots)\\ &= \Psi(\psi^k(a)),\end{aligned}$$

as desired. Finally, Ψ is injective because ψ^1 is the identity map. □

The following result due to [Knutson (1973)] is the main result of this section. It states that a $\mathbf{Z}$-torsion-free ψ-ring R induces a λ-ring structure on $R\otimes\mathbf{Q}$ by inverting the Newton Formulas (Theorem 3.10).

Theorem 3.49. *Let R be a $\mathbf{Z}$-torsion-free ψ-ring. Define operations*

$$\lambda^n\colon R\otimes\mathbf{Q}\to R\otimes\mathbf{Q}$$

for $n\geq 0$ by setting, for $a\in R\otimes\mathbf{Q}$, $\lambda^0(a)=1$ and

$$-t\frac{d}{dt}(\log\lambda_t(a))=\psi_{-t}(a). \tag{3.23}$$

Equipped with the operations λ^n, the ring $R\otimes\mathbf{Q}$ is a λ-ring.

As before, λ_t and ψ_{-t} are the power series $\lambda_t(a) = 1 + \sum_{n\geq 1} \lambda^n(a)t^n$ and $\psi_{-t}(a) = \sum_{n\geq 1}(-1)^n \psi^n(a)t^n$. The ring $R \otimes \mathbf{Q}$ is given the ψ-ring structure in statement (1) of Proposition 3.48.

Before giving the proof of this Theorem, we need some preliminary results.

Proposition 3.50. *With the same hypotheses as in Theorem 3.49, $R \otimes \mathbf{Q}$ is a pre-λ-ring in which $\lambda^n(1) = 0$ for $n \geq 2$.*

Proof. It suffices to establish axioms (2), (3), and (4) in Definition 1.10. That λ^1 is the identity map follows from the assumption that ψ^1 is the identity map. This proves axiom (2). For axiom (3), note that for $n \geq 1$, we have $\psi^n(1) = 1$. This implies that

$$\frac{d}{dt}(\log \lambda_t(1)) = 1 - t + t^2 - t^3 + \cdots = \frac{1}{1+t},$$

so we have $\lambda_t(1) = 1 + t$. This proves axiom (3).

Axiom (4) is equivalent to the condition

$$\lambda_t(a+b) = \lambda_t(a)\lambda_t(b). \tag{3.24}$$

Since each ψ^n is additive, so is ψ_{-t}. Therefore, we have

$$\begin{aligned} -t\frac{d}{dt}(\log \lambda_t(a+b)) &= \psi_{-t}(a+b) \\ &= \psi_{-t}(a) + \psi_{-t}(b) \\ &= -t\frac{d}{dt}(\log \lambda_t(a) + \log \lambda_t(b)) \\ &= -t\frac{d}{dt}(\log(\lambda_t(a)\lambda_t(b))). \end{aligned}$$

This implies the required condition (3.24). □

Proposition 3.51. *Let R be a $\mathbf{Z}$-torsion-free ring. Define a map*

$$L\colon \Lambda(R \otimes \mathbf{Q}) \to (R \otimes \mathbf{Q})^{\mathbf{N}} \tag{3.25}$$

by setting

$$L(1 + a_1 t + a_2 t^2 + \cdots) = (r_1, r_2, \ldots)$$

if and only if

$$-t\frac{d}{dt}(\log(1 + a_1 t + a_2 t^2 + \cdots)) = \sum_{n\geq 1}(-1)^n r_n t^n.$$

Then L is a ψ-isomorphism.

In this Proposition, the ψ-ring structure on $\Lambda(R \otimes \mathbf{Q})$ comes from the Adams operations on its universal λ-ring structure (Theorem 2.6). The ψ-ring structure on the Cartesian product ring $(R \otimes \mathbf{Q})^{\mathbf{N}}$ comes from Proposition 3.47.

Proof. First note that

$$L(1+at) = (a, a^2, a^3, \ldots), \tag{3.26}$$

since

$$\begin{aligned} -t\frac{d}{dt}(\log(1+at)) &= \frac{-at}{1+at} \\ &= -at + a^2t^2 - a^3t^3 + \cdots. \end{aligned}$$

Setting $a = 0$ and 1, respectively, we obtain

$$L(1) = (0, 0, 0, \ldots) \quad \text{and} \quad L(1+t) = (1, 1, 1, \ldots),$$

so L preserves the additive zero and the multiplicative identity.

To see that L is additive, recall that the addition $+_\Lambda$ (2.4) in the universal λ-ring $\Lambda(R \otimes \mathbf{Q})$ is given by multiplication of strict power series. For f and g in $\Lambda(R \otimes \mathbf{Q})$, we have

$$\begin{aligned} -t\frac{d}{dt}(\log(f +_\Lambda g)) &= -t\frac{d}{dt}(\log(f) + \log(g)) \\ &= -t\frac{d}{dt}(\log(f)) - t\frac{d}{dt}(\log(g)). \end{aligned}$$

This shows that L is additive.

Since L is additive, to check that it is multiplicative, it follows as in the proof of Theorem 2.6 that it suffices to consider 1-dimensional elements in $\Lambda(R \otimes \mathbf{Q})$. Given two 1-dimensional elements, $1 + at$ and $1 + bt$ in $\Lambda(R \otimes \mathbf{Q})$, their product (2.4) is $(1 + at) \times_\Lambda (1 + bt) = 1 + abt$. Therefore, it follows from (3.26) that

$$\begin{aligned} L((1+at) \times_\Lambda (1+bt)) &= L(1+abt) \\ &= (ab, (ab)^2, (ab)^3, \ldots) \\ &= (a, a^2, \ldots) \times (b, b^2, \ldots) \\ &= L(1+at) \times L(1+bt), \end{aligned}$$

where $\times$ is the (entry-wise) multiplication in $(R \otimes \mathbf{Q})^{\mathbf{N}}$. This shows that L is multiplicative.

So far we have shown that L is a ring homomorphism. To show that L commutes with ψ^n, again it suffices to consider 1-dimensional elements $1 + at$ in $\Lambda(R \otimes \mathbf{Q})$ because both L and ψ^n are additive. Since $\Lambda(R \otimes \mathbf{Q})$ is a λ-ring, ψ^n is the nth power map when applied to 1-dimensional elements (Proposition 3.5). Therefore, in $\Lambda(R \otimes \mathbf{Q})$ we have

$$\begin{aligned} \psi^n(1+at) &= \underbrace{(1+at) \times_\Lambda \cdots \times_\Lambda (1+at)}_{n \text{ times}} \\ &= 1 + a^n t. \end{aligned}$$

Using (3.26) and Proposition 3.47, we have

$$\begin{aligned} L(\psi^n(1+at)) &= L(1+a^n t) \\ &= (a^n, a^{2n}, a^{3n}, \ldots) \\ &= \psi^n((a, a^2, a^3, \ldots)) \\ &= \psi^n(L(1+at)). \end{aligned}$$

This shows that L is a ψ-homomorphism.

It remains to show that L is a bijection. The injectivity of L follows from the facts that it is additive and that its kernel is trivial. To see that L is surjective, suppose that $(r_1, r_2, \ldots)$ is an element in $(R \otimes \mathbf{Q})^{\mathbf{N}}$. The required elements a_n must satisfy

$$\frac{d}{dt}\left(\log(1 + a_1 t + a_2 t^2 + \cdots)\right) = \sum_{n \geq 1} (-1)^{n-1} r_n t^{n-1}.$$

From here we formally integrate and rewrite it in terms of exponentials:

$$1 + a_1 t + a_2 t^2 + \cdots = \exp\left(\sum_{n\geq 1} (-1)^{n-1} \frac{r_n}{n} t^n\right).$$

This equation allows us to write a_n in terms of $r_1, \ldots, r_n$ in $R \otimes \mathbf{Q}$, which shows that L is surjective.

This finishes the proof of the Proposition. □

Proof. [Proof of Theorem 3.49] By (1.8), (2.4), and Proposition 3.50 we have an injective map $\lambda_t \colon R \otimes \mathbf{Q} \to \Lambda(R \otimes \mathbf{Q})$ of additive groups that preserves the multiplicative identity. The diagram

$$\begin{array}{ccc} R \otimes \mathbf{Q} & \xrightarrow{\lambda_t} & \Lambda(R \otimes \mathbf{Q}) \\ & \searrow^{\Psi} & \cong \downarrow L \\ & & (R \otimes \mathbf{Q})^{\mathbf{N}} \end{array}$$

is commutative because of the definitions of Ψ (3.22), λ_t (3.23), and L (3.25). The map Ψ is an injective ψ-homomorphism by Proposition 3.48, and the map L is a ψ-isomorphism by Proposition 3.51.

It follows that λ_t is an injective ψ-homomorphism as well. In particular, λ_t is a ring homomorphism, and it commutes with all the operations ψ^k. By Theorem 3.15 each operation λ^n on $R \otimes \mathbf{Q}$ (and $\Lambda(R \otimes \mathbf{Q})$) is a polynomial with rational coefficients in $\psi^1, \ldots, \psi^n$. Therefore, we conclude that λ_t commutes with each λ^n. In other words, λ_t is a pre-λ-homomorphism. By Proposition 2.8 this implies that $R \otimes \mathbf{Q}$ is a λ-ring, as desired. □

3.5 ψ-Rings as Ψ-Coalgebras

In this section, we describe the comonad Ψ on the category of rings whose coalgebras are exactly the ψ-rings following [Patras (2003)].

For a ring R, let

$$\Psi(R) = R^{\mathbf{N}} = \prod_{i=0}^{\infty} R$$

be the Cartesian product ring (3.20), whose ring operations are defined entry-wise. If $f\colon R \to S$ is a ring homomorphism, let $\Psi(f)\colon \Psi(R) \to \Psi(S)$ be the map defined by $\Psi(f)(a_1, a_2, \ldots) = (f(a_1), f(a_2), \ldots)$. Define a map $\delta\colon \Psi(R) \to \Psi(\Psi(R))$ by

$$\delta(\mathbf{a}) = (\psi^1(\mathbf{a}), \psi^2(\mathbf{a}), \psi^3(\mathbf{a}), \ldots)$$

for $\mathbf{a} = (a_1, a_2, \ldots) \in \Psi(R)$, where the ψ^k are the ones in Proposition 3.47. Finally, define a map $\nu\colon \Psi(R) \to R$ by $\nu(\mathbf{a}) = a_1$.

Theorem 3.52. *With the above definitions, (Ψ, δ, ν) is a comonad on the category of rings (Definition 2.19).*

Proof. It is clear that $\Psi(f)$ is a ring homomorphism because the ring operations are defined entry-wise. It is also clear that Ψ preserves the identity maps and compositions. Thus, Ψ is a functor on the category of rings.

To see that $\delta\colon \Psi \to \Psi^2$ is a natural transformation, first observe that, when applied to a ring R, it is a ring homomorphism. This is because $\Psi(R)$ is a ψ-ring (Proposition 3.47) and the ring operations in $\Psi^2(R)$ are defined entry-wise. Now let $f\colon R \to S$ be a ring homomorphism, and let $\mathbf{a}$ be an element in $\Psi(R)$. Direct inspection shows that both $\Psi^2(f)(\delta(\mathbf{a}))$ and $\delta(\Psi(f)(\mathbf{a}))$ are equal to

$$((f(a_1), f(a_2), \ldots), (f(a_2), f(a_4), \ldots), (f(a_3), f(a_6), \ldots), \ldots).$$

In other words, the square

$$\begin{array}{ccc} \Psi(R) & \xrightarrow{\delta} & \Psi^2(R) \\ {\scriptstyle \Psi(f)}\downarrow & & \downarrow{\scriptstyle \Psi^2(f)} \\ \Psi(S) & \xrightarrow{\delta} & \Psi^2(S) \end{array}$$

is commutative, showing that δ is a natural transformation.

Next, we have that $\nu\colon \Psi \to Id$ is a natural transformation, which follows immediately from the definitions.

To show that (Ψ, δ, ν) is a comonad, we need to show that the two diagrams (2.21) and (2.22) commute in the present setting. The commutativity of the square

$$\begin{array}{ccc} \Psi(R) & \xrightarrow{\delta} & \Psi^2(R) \\ {\scriptstyle \delta}\downarrow & & \downarrow{\scriptstyle \Psi\delta} \\ \Psi^2(R) & \xrightarrow{\delta\Psi} & \Psi^3(R) \end{array}$$

follows from the observation that both $(\Psi\delta)(\delta(\mathbf{a}))$ and $(\delta\Psi)(\delta(\mathbf{a}))$ are equal to

$$\Big(\ldots, \underbrace{(\psi^n(\mathbf{a}), \psi^{2n}(\mathbf{a}), \psi^{3n}(\mathbf{a}), \ldots)}_{n^{\text{th}} \text{ entry}}, \ldots\Big).$$

The commutativity of the diagram

$$\begin{array}{ccccc} & & \Psi(R) & & \\ & /\!\!/ & \downarrow{\scriptstyle \delta} & \backslash\!\!\backslash & \\ \Psi(R) & \xleftarrow{\nu\Psi} & \Psi^2(R) & \xrightarrow{\Psi\nu} & \Psi(R) \end{array} \tag{3.27}$$

can be checked similarly. We leave the details to the reader as an exercise. □

Now we identify the coalgebras (Definition 2.21) of the comonad (Ψ, δ, ν) with ψ-rings.

Theorem 3.53. *The Ψ-coalgebras are exactly the ψ-rings (Definition 3.44).*

Proof. Let R be a Ψ-coalgebra with structure map $\rho\colon R \to \Psi(R)$. Then we set $\rho(a) = (\psi^1(a), \psi^2(a), \psi^3(a), \ldots)$ for $a \in R$. To show that R is a ψ-ring when equipped with these operations ψ^k, we need to show that each ψ^n is a ring homomorphism, that ψ^1 is the identity map, and that $\psi^m\psi^m$ is equal to ψ^{mn}. Since ρ is a ring homomorphism and since the ring operations in $\Psi(R)$ are defined entry-wise, it follows that each ψ^n is also a ring homomorphism.

The condition $\nu\rho = Id$ for a Ψ-coalgebra says that

$$\begin{aligned} a &= \nu\rho(a) \\ &= \nu\left(\psi^1(a), \psi^2(a), \psi^3(a), \ldots\right) \\ &= \psi^1(a) \end{aligned}$$

for $a \in R$, so ψ^1 is the identity map.

From the commutativity of the square

$$\begin{array}{ccc} R & \xrightarrow{\rho} & \Psi(R) \\ {\scriptstyle \rho}\downarrow & & \downarrow{\scriptstyle \delta} \\ \Psi(R) & \xrightarrow{\Psi(\rho)} & \Psi^2(R), \end{array}$$

we see that

$$\delta(\rho(a)) = \left(\ldots, \left(\psi^n(a), \psi^{2n}(a), \psi^{3n}(a), \ldots\right), \ldots\right)$$

is equal to

$$\Psi(\rho)(\rho(a)) = \left(\ldots, \left(\psi^1\psi^n(a), \psi^2\psi^n(a), \psi^3\psi^n(a), \ldots\right), \ldots\right)$$

for $a \in R$. Therefore, we have $\psi^m\psi^n = \psi^{mn}$ for all $m, n \geq 1$. We have shown that R is a ψ-ring.

For the converse, simply reverse the above arguments. □

3.6 Wilkerson's Theorem

Recall from Theorem 3.49 that a **Z**-torsion-free ψ-ring (Definition 3.44) induces a λ-ring structure on $R \otimes \mathbf{Q}$ by inverting the Newton Formulas. The main purpose of this section is to refine this result to a λ-ring structure on R itself. This refinement is due to [Wilkerson (1982)], and it is a very useful tool for constructing λ-rings, as we will see in the next section and later chapters.

There is an additional assumption in order to obtain λ-operations on R itself. Recall that when p is a prime, the Adams operation ψ^p in a λ-ring R satisfies (Proposition 3.4) $\psi^p(a) \equiv a^p \pmod{pR}$ for $a \in R$. With this additional assumption on a **Z**-torsion-free ψ-ring, the λ-operations on $R \otimes \mathbf{Q}$ actually restrict to maps on R itself. This gives the required λ-ring structure.

Here is the main result of this section.

Theorem 3.54 (Wilkerson's Theorem). *Let R be a **Z**-torsion-free ψ-ring in which*

$$\psi^p(x) \equiv x^p \pmod{pR} \tag{3.28}$$

for each $x \in R$ and prime p. Then there exists a unique λ-ring structure on R whose Adams operations are the ones from the given ψ-ring structure.

In particular, Wilkerson's Theorem says that, in order to construct a λ-ring structure on a **Z**-torsion-free ring R, it suffices to construct the Adams operations, i.e., a ψ-ring structure that satisfies (3.28).

To prove Wilkerson's Theorem, we first need some preliminary results.

Lemma 3.55 ([Hazewinkel (1978)] Sublemma (17.6.6)). *Let R be a **Z**-torsion-free ring, and let $x \in R \otimes \mathbf{Q}$. If $x \in R \otimes \mathbf{Z}_{(p)}$ for every prime p, then $x \in R$.*

Proof. Since $x \in R \otimes \mathbf{Q}$, we can write $x = n^{-1}y$ for some $y \in R$ and integer $n \geq 1$. If $n = 1$, then we are done. If $n > 1$, then write it as a product $n = pm$ for some prime p and integer $m \geq 1$. Since $x \in R \otimes \mathbf{Z}_{(p)}$, we can also write $x = k^{-1}z$ for some $z \in R$ and integer $k \geq 1$ that is relatively prime to p. In particular, there are integers a and b such that $ap + bk = 1$. We know that $nkx = ky = nz$, so we have

$$\begin{aligned} y &= apy + bky \\ &= apy + bnz \\ &= p(ay + bmz) \in pR. \end{aligned}$$

In other words, we can write $x = m^{-1}y'$ in which $y' \in R$ and $1 \leq m < n$. By a downward induction it follows that x must lie in R itself. □

Lemma 3.56. *In any λ-ring R, the following statements hold for $x \in R$, prime p, and $k, l \geq 1$:*

$$\begin{aligned} \psi^p(x) - x^p &= p\left\{(-1)^{p+1}\lambda^p(x) + \alpha(\lambda^1(x), \ldots, \lambda^{p-1}(x))\right\}, \\ \lambda^k(\lambda^l(x)) &= (-1)^{(k+1)(l+1)}\lambda^{kl}(x) + \beta(\lambda^1(x), \ldots, \lambda^{kl-1}(x)). \end{aligned} \tag{3.29}$$

Here α and β are some polynomials with integer coefficients.

Proof. By Theorem 1.39 and The Verification Principle (Theorem 1.43), it suffices to consider the case $R = U = \mathbf{Z}[s_1, s_2, s_3, \ldots]$, the free λ-ring on one generator (Proposition 1.38), and $x = s_1 \in U$.

For the first assertion, from the definition of Adams operations (3.1) we have

$$\begin{aligned} &\psi_{-t}(x) \\ &= -t(\lambda_t(x))' \times (\lambda_t(x))^{-1} \\ &= \left(-xt - 2\lambda^2(x)t^2 - \cdots - n\lambda^n(x)t^n - \cdots\right) \times \\ &\quad \left(\cdots + ((-1)^{n-1}x^{n-1} + (\text{polynomial in } \lambda^i(x), i < n))t^{n-1} + \cdots\right) \\ &= \cdots + \left((-1)^n x^n - n\lambda^n(x) + (\text{polynomial in } \lambda^i(x), i < n)\right) t^n + \cdots. \end{aligned}$$

Thus, we have

$$\psi^n(x) = x^n + (-1)^{n+1}n\lambda^n(x) + (\text{polynomial in } \lambda^i(x), i < n).$$

The first assertion follows from this when $n = p$ and the fact that $\psi^p(x) \equiv x^p \pmod{pR}$ in a λ-ring (Proposition 3.4).

For the second assertion, we use the Newton Formula (Theorem 3.10) and Theorem 3.9 to infer that

$$\begin{aligned} \psi^{kl}(x) &= (-1)^{kl+1}kl\lambda^{kl}(x) + \lambda^1(x)\psi^{kl-1}(x) + \cdots + (-1)^{kl}\lambda^{kl-1}(x)\psi^1(x) \\ &= (-1)^{kl+1}kl\lambda^{kl}(x) + (\text{polynomial in } \lambda^i(x), i < kl). \end{aligned}$$

Likewise, we have

$$\begin{aligned} \psi^k(\psi^l(x)) &= (-1)^{k+1}k\lambda^k\left((-1)^{l+1}l\lambda^l(x) + (\text{polynomial in } \lambda^i(x), i < l)\right) \\ &\quad + (\text{polynomial in } \lambda^i(x), i < kl) \\ &= (-1)^{k+1}k\lambda^k\left((-1)^{l+1}l\lambda^l(x)\right) + (\text{polynomial in } \lambda^i(x), i < kl) \\ &= (-1)^{(k+1)+(l+1)}kl\lambda^k(\lambda^l(x)) + (\text{polynomial in } \lambda^i(x), i < kl). \end{aligned}$$

These equalities follow from the λ-ring axioms for $\lambda^n(a+b)$, $\lambda^n(ab)$, and $\lambda^n(\lambda^n(a))$ (Definition 1.10) and Proposition 1.14. Since $\psi^{kl} = \psi^k\psi^l$ in any λ-ring (3.2), it follows that

$$(-1)^{k+l}kl\lambda^k(\lambda^l(x)) = (-1)^{kl+1}kl\lambda^{kl}(x) + \beta', \tag{3.30}$$

where β' is some polynomial with integer coefficients in $\lambda^i(x)$ for $i < kl$. Since we are working in the free λ-ring U on one generator with $x = s_1$, both the left-hand side and the first term on the right-hand side of (3.30) are polynomials with integer coefficients in the $s_j \in U$, and both are divisible by kl. Therefore, since U is the polynomial ring over $\mathbf{Z}$ in the s_j $(j \geq 1)$, it follows that the coefficients in β' are all divisible by kl. In other words, β' has the form

$$\beta' = kl\beta(\lambda^1(x), \ldots, \lambda^{kl-1}(x))$$

for some polynomial β with integer coefficients in $\lambda^i(x)$ for $i < kl$. The second assertion of the Lemma now follows from (3.30) by dividing by $(-1)^{k+l}kl$. □

We are now ready to prove Wilkerson's Theorem.

Proof. [Proof of Theorem 3.54] The uniqueness assertion follows from the fact that in a $\mathbf{Z}$-torsion-free λ-ring, the λ-ring structure is completely determined by its Adams operations (Theorem 3.15).

For the existence part, from Theorem 3.49 we already know that $R \otimes \mathbf{Q}$ is a λ-ring in which the Adams operations are the given ψ^k. Thus, by Lemma 3.55 it suffices to show that the λ-operations on $R \otimes \mathbf{Q}$ satisfies

$$\lambda^n \colon R \otimes \mathbf{Z}_{(p)} \to R \otimes \mathbf{Z}_{(p)} \tag{3.31}$$

for each prime p. This is done by induction on n. The case $n = 1$ is trivially true, since λ^1 is the identity map.

Note that each ψ^k maps $R \otimes \mathbf{Z}_{(p)}$ to itself, since ψ^k is actually defined on R and it is a ring homomorphism. Moreover, it follows from the assumption (3.28) that we have $\psi^p(x) \equiv x^p \pmod{p(R \otimes \mathbf{Z}_{(p)})}$ for $x \in R \otimes \mathbf{Z}_{(p)}$.

Pick an element $x \in R \otimes \mathbf{Z}_{(p)}$ for the rest of this proof. Suppose that $n < p$. In the Newton Formula (Theorem 3.10)

$$\sum_{i=0}^{n-1} (-1)^i \lambda^i(x) \psi^{n-i}(x) = (-1)^{n+1} n \lambda^n(x),$$

every summand $\lambda^i(x)\psi^{n-i}(x)$ on the left-hand side lies in $R \otimes \mathbf{Z}_{(p)}$ by the induction hypothesis, hence so does $n\lambda^n(x)$. Since $n < p$, it is also relatively prime to p, so $n^{-1} \in \mathbf{Z}_{(p)}$. This shows that (3.31) is true for $n < p$.

For $n = p$ we have

$$\psi^p(x) - x^p = p\left\{(-1)^{p+1}\lambda^p(x) + \alpha(\lambda^1(x), \ldots, \lambda^{p-1}(x))\right\}$$

by Lemma 3.56. Since $\psi^p(x) - x^p \in p(R \otimes \mathbf{Z}_{(p)})$, it follows that

$$(-1)^{p+1}\lambda^p(x) + \alpha(\lambda^1(x), \ldots, \lambda^{p-1}(x)) \in R \otimes \mathbf{Z}_{(p)}.$$

But we have already shown that $\lambda^i(x) \in R \otimes \mathbf{Z}_{(p)}$ for $i < p$, so

$$\alpha(\lambda^1(x), \ldots, \lambda^{p-1}(x)) \in R \otimes \mathbf{Z}_{(p)}$$

as well. This implies that $\lambda^p(x)$ lies in $R \otimes \mathbf{Z}_{(p)}$.

For $n > p$ first suppose that n is relatively prime to p. In this case, we can use the Newton Formula exactly as we did two paragraphs ago. Since n is invertible in $\mathbf{Z}_{(p)}$, (3.31) is true in this case.

On the other hand, suppose that $n = kp$ for some integer $k > 1$. By Lemma 3.56 we know that

$$\lambda^n(x) = (-1)^{(p+1)(k+1)}\lambda^k(\lambda^p(x)) \pm \beta(\lambda^1(x), \ldots, \lambda^{n-1}(x)).$$

We already know that $\lambda^p(x) \in R \otimes \mathbf{Z}_{(p)}$, hence so does $\lambda^k(\lambda^p(x))$ by the induction hypothesis, since $k < n$. The induction hypothesis also implies that $\beta(\lambda^1(x), \ldots, \lambda^{n-1}(x)) \in R \otimes \mathbf{Z}_{(p)}$. Therefore, we have $\lambda^n(x) \in R \otimes \mathbf{Z}_{(p)}$, as desired. This finishes the proof of Wilkerson's Theorem 3.54. □

3.7 Localization and Completion of λ-Rings

The purpose of this section is to show that λ-rings are closed under most reasonable localizations (Theorem 3.58) and completions (Theorem 3.62). These results are applications of Wilkerson's Theorem 3.54 and can be found in [Wilkerson (1982)]. A general reference for localizations and completions of rings is [Sullivan (2005)].

Applied to the λ-ring $\mathbf{Z}$ of integers, it follows that the p-local integers $\mathbf{Z}_{(p)}$ (Corollary 3.59) and the p-adic integers $\hat{\mathbf{Z}}_p$ (Corollary 3.63) are both λ-rings in which the Adams operations are all equal to the identity.

We begin with localizations.

3.7.1 *Localization of λ-rings*

An element $a \neq 0$ in a ring R is called a *zero-divisor* if there exists an element $b \neq 0$ in R such that $ab = 0$. Note that a characteristic 0 ring R without zero-divisors is a $\mathbf{Z}$-torsion-free ring (Definition 3.12). In particular, a λ-ring without zero-divisors must be $\mathbf{Z}$-torsion-free because λ-rings have characteristic 0 (Proposition 1.29).

Let R be a ring without zero-divisors. A *multiplicative subset* S of R is a subset $S \subseteq R - \{0\}$ containing the multiplicative identity 1 of R such that, if a and b are both in S, then so is ab. The *localization* $S^{-1}R$ is the ring (also without zero-divisors) whose elements are equivalence classes of "quotients" x/s with $x \in R$ and $s \in S$. The equivalence relation is given by

$$x/s \sim y/r \quad \text{if and only if} \quad rx = sy.$$

The ring operations in $S^{-1}R$ are given by the usual addition and multiplication of quotients. In other words, in terms of representatives, we have

$$\begin{aligned}(x/s) + (y/r) &= (rx + sy)/(sr),\\ (x/s) \cdot (y/r) &= (xy)/(sr).\end{aligned}$$

There is an injective ring homomorphism $i\colon R \to S^{-1}R$ defined by $i(r) = r/1$. Here we write x/s for the element in $S^{-1}R$ represented by it.

Example 3.57. In the ring $\mathbf{Z}$ of integers, let p be a prime, and let S_p be the multiplicative subset of $\mathbf{Z}$ consisting of the integers that are *not* divisible by p.

Then the localization $S_p^{-1}\mathbf{Z} = \mathbf{Z}_{(p)}$, which is the ring of p-local integers. In other words, $\mathbf{Z}_{(p)}$ is obtained from the integers by inverting all the primes that are not equal to p.

Instead of taking S_p, one can take any multiplicative subset I of $\mathbf{Z}$ generated by a set of primes. Then the localization (away from I) $I^{-1}\mathbf{Z}$ is obtained from $\mathbf{Z}$ by inverting all the primes in I. For example, to localize $\mathbf{Z}$ at the primes 2 and 3, we take the multiplicative subset I generated by the primes $p > 3$. Then the localization $I^{-1}\mathbf{Z} = \mathbf{Z}_{(2,3)}$ is the localization of $\mathbf{Z}$ at 2 and 3.

Here is one useful consequence of Wilkerson's Theorem.

Theorem 3.58. *Let R be a λ-ring without zero-divisors, and let S be a multiplicative subset of R such that $\psi^k(S) \subseteq S$ for $k \geq 1$. Then the localization $S^{-1}R$ has a λ-ring structure that extends the one on R in the sense that $\psi^k(x/s) = \psi^k(x)/\psi^k(s)$ for $x \in R$, $s \in S$, and $k \geq 1$.*

Proof. For each integer $k \geq 1$, the function $\psi^k \colon S^{-1}R \to S^{-1}R$ defined by $\psi^k(x/s) = \psi^k(x)/\psi^k(s)$ is easily seen to be well-defined and is a ring homomorphism. Moreover, it is clear that these extended maps ψ^k satisfy $\psi^1 = Id$ and $\psi^n\psi^m = \psi^{nm}$, since they do on R. Thus, the localization $S^{-1}R$ is a ψ-ring. It is also $\mathbf{Z}$-torsion-free because it has no zero-divisors and has characteristic 0. Finally, if p is a prime, then we have

$$\begin{aligned}\psi^p(x/s) - (x/s)^p &= \psi^p(x)/\psi^p(s) - x^p/s^p \\ &= \left\{s^p(\psi^p(x) - x^p) + (s^p - \psi^p(s))x^p\right\}/(s^p\psi^p(s)).\end{aligned}$$

Since both $\psi^p(x) - x^p$ and $s^p - \psi^p(s)$ lie in pR, it follows that $\psi^p(x/s) - (x/s)^p \in p(S^{-1}R)$. Therefore, Wilkerson's Theorem 3.54 applies to $S^{-1}R$, and we obtain the desired λ-ring structure on $S^{-1}R$. □

The following result is an immediate consequence of the above Theorem.

Corollary 3.59. *Any localization of the integers $\mathbf{Z}$ at a set of primes is a λ-ring with $\psi^k = Id$ for all k. In particular, the p-local integers $\mathbf{Z}_{(p)}$ form such a λ-ring.*

Proof. Apply Theorem 3.58 to the λ-ring $\mathbf{Z}$, which satisfies $\psi^k = Id$ for all k (Example 3.4). □

3.7.2 *Completions of λ-rings*

First we recall a few definitions from real analysis.

A *metric* on a non-empty set X is a real-valued function $\rho \colon X \times X \to \mathbf{R}$ such that for $x, y, z \in X$:

(1) $\rho(x, y) \geq 0$.
(2) $\rho(x, y) = 0$ if and only if $x = y$.
(3) $\rho(x, y) = \rho(y, x)$.

(4) $\rho(x, z) \leq \rho(x, y) + \rho(y, z)$.

Such a metric makes X into a topological space, called a *metric space*. The reader can consult, for example, [Munkres (1975); Royden (1988)] for more discussion of metric spaces and their completions.

A *topological ring* is a ring R together with a metric ρ on R such that the ring operations $+, \times : R \times R \to R$ on R are continuous with respect to the metric topology.

Example 3.60. Fix a prime p. Define a real-valued function $\rho\colon \mathbf{Z} \times \mathbf{Z} \to \mathbf{R}$ by setting $\rho(x, y) = 2^{-k}$ if and only if $x - y \equiv 0 \pmod{p^k}$ but $x - y \not\equiv 0 \pmod{p^{k+1}}$. Note that $\rho(x, y) = 0$ if and only if $x = y$. The other axioms for a metric are also easy to verify. This metric ρ is called the *p-adic metric* on $\mathbf{Z}$. With this metric topology, the integers $\mathbf{Z}$ becomes a topological ring. (The reader should check this.)

A *convergent sequence* in a metric space (X, ρ) is a sequence $\{x_n\}$ of elements in X such that there exists an element $x \in X$ and for ever $\epsilon > 0$, there exists an integer $N > 0$ such that $\rho(x_n, x) < \epsilon$ whenever $n > N$. A *Cauchy sequence* in a metric space (X, ρ) is a sequence $\{x_n\}$ of elements in X such that for ever $\epsilon > 0$, there exists an integer $N > 0$ such that $\rho(x_n, x_m) < \epsilon$ whenever $n, m > N$. A convergent sequence is a Cauchy sequence, but not every Cauchy sequence in a metric space is convergent. A metric space is said to be *complete* if every Cauchy sequence is convergent.

If a metric space (X, ρ) is not complete, there is a canonical way to associate to it a complete metric space X_ρ, called the *ρ-completion* of X. The points in X_ρ are Cauchy sequences in X, and two Cauchy sequences $\{x_n\}$ and $\{y_n\}$ are identified in X_ρ if

$$\rho^*(\{x_n\}, \{y_n\}) = \lim \rho(x_n, y_n) = 0.$$

Then (X_ρ, ρ^*) is a complete metric space, and X is isometrically embedded in X_ρ as a dense subset by sending a point $x \in R$ to the Cauchy sequence $\{x, x, \ldots\}$.

Moreover, if (X, ρ) is a topological ring, then so is the ρ-completion X_ρ. The ring operations in X_ρ are defined component-wise:

$$\{x_n\} + \{y_n\} = \{x_n + y_n\},$$
$$\{x_n\} \cdot \{y_n\} = \{x_n y_n\}.$$

Example 3.61. In the context of Example 3.60, the ρ-completion of $\mathbf{Z}$ is denoted by $\hat{\mathbf{Z}}_p$ and is called the ring of *p-adic integers*. A point in $\hat{\mathbf{Z}}_p$ is represented by a sequence $\{x_n\}$ of integers such that for every integer $k \geq 1$, there exists an integer $N \geq 1$ such that $x_n - x_m \equiv 0 \pmod{p^k}$ whenever $n, m > N$.

Theorem 3.62. *Let R be a $\mathbf{Z}$-torsion-free λ-ring, and let ρ be a metric on R with respect to which R is a topological ring. Suppose that each Adams operation ψ^k in R maps Cauchy sequences to Cauchy sequences. Then the ρ-completion R_ρ has a λ-ring structure that extends the one on R.*

Proof. We leave it as an exercise to show that the ρ-completion R_ρ is $\mathbf{Z}$-torsion-free and that the function $\psi^k\colon R_\rho \to R_\rho$ defined by $\psi^k(\{x_n\}) = \{\psi^k x_n\}$ is well-defined, where $\{x_n\}$ is a Cauchy sequence. Each ψ^k on R_ρ is a ring homomorphism because it is so on R. Moreover, it is immediate that $\psi^1 = Id$ and $\psi^n\psi^m = \psi^{nm}$ on R_ρ. Thus, R_ρ is a $\mathbf{Z}$-torsion-free ψ-ring. If $\{x_n\}$ is a Cauchy sequence and p is a prime, then $\psi^p(\{x_n\}) - (\{x_n\})^p = \{\psi^p(x_n) - x_n^p\}$. Since each $\psi^p(x_n) - x_n^p \in pR$, it follows that $\psi^p(\{x_n\}) - (\{x_n\})^p \in pR_\rho$. Therefore, Wilkerson's Theorem 3.54 applies to R_ρ, giving the desired λ-ring structure. □

Corollary 3.63. *The ring $\hat{\mathbf{Z}}_p$ of p-adic integers (Example 3.61) is a λ-ring in which every Adams operation is equal to the identity map.*

Proof. Apply Theorem 3.62 to the λ-ring $\mathbf{Z}$ with the p-adic metric (Example 3.60). Since $\psi^k = Id$ on $\mathbf{Z}$ for all k, the same is true on $\hat{\mathbf{Z}}_p$. □

3.8 Exercises

(1) Prove Propositions 3.35 and 3.37 and Theorems 3.38 , 3.39, and 3.40. Prove their analogs for natural operations on γ-rings with $r \geq 1$ inputs.
(2) Which of the properties of Adams operations and γ-operations that are proved in this Chapter still hold in a pre-λ-ring (Definition 1.10)?
(3) Write down explicitly the polynomial $G_{n,m}$ in Corollary 3.24.
(4) Check that $\mathbf{Z}$ is a topological ring with respect to the p-adic metric ρ in Example 3.60.
(5) Show that $\mathbf{Z}$ is *not* complete with respect to the p-adic metric ρ in Example 3.60.
(6) Prove that there exists a canonical isomorphism from the ring $\hat{\mathbf{Z}}_p$ of p-adic integers to the inverse limit $\varprojlim_n \mathbf{Z}/p^n$ of the inverse system
$$\mathbf{Z}/p \leftarrow \mathbf{Z}/p^2 \leftarrow \mathbf{Z}/p^3 \leftarrow \cdots.$$
(7) Let S be a non-empty set, and let $\mathbf{k}$ be a field of characteristic 0. Denote by $\mathbf{k}(S)$ the set of functions from S to $\mathbf{k}$.
   (a) Show that $\mathbf{k}(S)$ is a ring with the usual addition and multiplication of functions.
   (b) Show that $\mathbf{k}(S)$ has a subring isomorphic to $\mathbf{Q}$.
   (c) Suppose that S comes with a sequence of functions $\phi_n\colon S \to S$ for $n \geq 1$ such that
      - ϕ_1 is the identity function, and
      - $\phi_n\phi_m = \phi_{nm}$ for all $n, m \geq 1$.

   For $n \geq 1$ define a function $\psi^n\colon \mathbf{k}(S) \to \mathbf{k}(S)$ by setting $\psi^n(f)(s) = f(\phi_n(s))$ for $f \in \mathbf{k}(S)$ and $s \in S$. Prove: $\mathbf{k}(S)$ is a λ-ring with Adams operations ψ^n.

(d) Let G be a group, and let C_G be the set of conjugacy classes of G. Let ϕ_n be the function on G defined by $\phi_n(g) = g^n$ for $g \in G$ and $n \geq 1$. Prove: ϕ_n induces a function on C_G, and these induced functions satisfy the hypotheses of part (7c). So $\mathbf{k}(C_G)$ is a λ-ring, called the *λ-ring of central functions on G with values in* $\mathbf{k}$.

Chapter 4

Witt Vectors

The purposes of this Chapter are (1) to discuss the ring $\mathbf{W}(R)$ of big Witt vectors on R and (2) to describe the relationships between the universal λ-ring $\Lambda(R)$ and the ring $\mathbf{W}(R)$. Witt vectors play an important role in formal group theory. Indeed, the ring $\mathbf{W}(R)$ is closely related to the group of curves over a formal group law. We will not discuss formal group laws in this book. Instead, we will construct the ring $\mathbf{W}(R)$ and show that it is isomorphic to the universal λ-ring. In fact, there is a natural isomorphism $E\colon \Lambda \to \mathbf{W}$, called the *Artin-Hasse Exponential isomorphism*, between these functors on the category of rings (Theorem 4.16). Using the Artin-Hasse Exponential isomorphism, the comonad structure on the functor Λ can be pulled over to $\mathbf{W}$, making it into a comonad. We will describe the comonad $\mathbf{W}$ as well.

In §4.1 the ring $\mathbf{W}(R)$ of big Witt vectors on R is constructed using Dwork's Theorem 4.4. In §4.2 we discuss the Artin-Hasse Exponential isomorphism $E\colon \Lambda \to \mathbf{W}$ of functors on the category of rings (Theorem 4.16). In §4.3 we describe the analogs of the Witt polynomials w_n for the universal λ-ring functor Λ (Theorem 4.21) and use them to give a characterization of Λ (Corollary 4.22). In §4.4 we characterize the comonad Δ on the big Witt vectors functor $\mathbf{W}$. At the end of §4.4, we provide a dictionary between the functors Λ and $\mathbf{W}$ and their associated natural transformations.

Most of the materials in this section can be found in Section 17 in [Hazewinkel (1978)]. In defining the ring structure in $\mathbf{W}(R)$, the key result is Dwork's Theorem 4.4, which originated in the work of Dieudonné [Dieudonné (1957)] and Dwork [Dwork (1958)]. For discussion about the use of Witt vectors in formal group theory, the reader is referred to [Dieudonné (1973); Hazewinkel (1978); Lenart (1998)]. Combinatorial descriptions and generalizations of Witt vectors can be found in [Dress and Siebeneicher (1988); Graham (1993); Metropolis and Rota (1983)].

4.1 Big Witt Vectors

Given a ring R, the ring $\mathbf{W}(R)$ of big Witt vectors on R is defined in terms of a sequence of polynomials w_n, called the Witt polynomials. The key result is Dwork's Theorem 4.4, which allows us to unambiguously define the ring operations on $\mathbf{W}(R)$ for any ring R.

4.1.1 *The Witt polynomials*

Let R be a ring. Define the underlying set of $\mathbf{W}(R)$ as the infinite product $\mathbf{W}(R) = \prod_{i=1}^{\infty} R$. An element in $\mathbf{W}(R)$ is denoted by $\mathbf{a} = (a_1, a_2, \ldots)$ for $a_i \in R$.

Let $\mathbf{x} = (x_1, x_2, \ldots)$ be a countably infinite sequence of variables. For $n \geq 1$ define the polynomial with integer coefficients,

$$w_n(\mathbf{x}) = w_n(x_1, \ldots, x_n) = \sum_{d|n} d x_d^{\frac{n}{d}}, \tag{4.1}$$

where the sum is taken over all the positive integers d that divide n. The ring operations in $\mathbf{W}(R)$ are defined using these polynomials w_n, called the *Witt polynomials.*

Example 4.1. Here are the first few Witt polynomials and w_p for p prime:

$$\begin{aligned}
w_1(\mathbf{x}) &= x_1, \\
w_2(\mathbf{x}) &= x_1^2 + 2x_2, \\
w_3(\mathbf{x}) &= x_1^3 + 3x_3, \\
w_4(\mathbf{x}) &= x_1^4 + 2x_2^2 + 4x_4, \\
w_5(\mathbf{x}) &= x_1^5 + 5x_5, \\
w_6(\mathbf{x}) &= x_1^6 + 2x_2^3 + 3x_3^2 + 6x_6, \\
w_p(\mathbf{x}) &= x_1^p + p x_p \quad (p \text{ prime}).
\end{aligned}$$

4.1.2 *The Witt addition and multiplication*

The Witt polynomials can be applied to elements in $\mathbf{W}(R)$, giving rise to functions $w_n \colon \mathbf{W}(R) \to R$, where $w_n(\mathbf{a}) = \sum_{d|n} d a_d^{\frac{n}{d}}$. The Witt addition $+_W$ and multiplication $\times_W$ in $\mathbf{W}(R)$ will be defined in such a way that the functions w_n are ring homomorphisms:

$$\begin{aligned}
w_n(\mathbf{a} +_W \mathbf{b}) &= w_n(\mathbf{a}) + w_n(\mathbf{b}), \\
w_n(\mathbf{a} \times_W \mathbf{b}) &= w_n(\mathbf{a}) \cdot w_n(\mathbf{b}).
\end{aligned} \tag{4.2}$$

The actual definition will be given in Definition 4.9 after some preliminary work.

One can think of $w_n(\mathbf{a})$ as the "nth digit" of $\mathbf{a}$. So the Witt addition $\mathbf{a} +_W \mathbf{b}$ is defined so that its nth digit is the sum of the nth digits of $\mathbf{a}$ and $\mathbf{b}$. The same remark goes for the Witt multiplication.

It is not entirely straightforward to see why the elements $\mathbf{a} +_W \mathbf{b}$ and $\mathbf{a} \times_W \mathbf{b}$ should exist that satisfy (4.2). Even if they do exist, we still need to make sure that they are unique.

Proposition 4.2. *Let R be a $\mathbf{Q}$-algebra, and let $\mathbf{a}$ and $\mathbf{b}$ be elements in $\mathbf{W}(R)$. Then there exist unique elements $\mathbf{a} +_W \mathbf{b}$ and $\mathbf{a} \times_W \mathbf{b}$ in $\mathbf{W}(R)$ such that the two conditions in* (4.2) *are satisfied.*

Proof. It suffices to show that, when R is a $\mathbf{Q}$-algebra, one can write x_n uniquely in terms of the polynomials $w_1, \ldots, w_n$ with rational coefficients. This is clear when $n = 1$, since $w_1 = x_1$. Suppose that for $k \leq n$, x_k can be written uniquely as a polynomial in $w_1, \ldots, w_k$ with rational coefficients. Then we have

$$(n+1)x_{n+1} = w_{n+1} - \sum_{\substack{d|(n+1) \\ d \leq n}} d x_d^{\frac{n+1}{d}}.$$

The induction hypothesis applies to the terms x_d on the right-hand side of this equation. Now divide both sides of the above equality by $n+1$, which gives

$$x_{n+1} = \frac{1}{n+1} w_{n+1} - \frac{1}{n+1} \sum_{\substack{d|(n+1) \\ d \leq n}} d x_d^{\frac{n+1}{d}}.$$

The right-hand side of this equality is a rational polynomial in $w_1, \ldots, w_{n+1}$. The uniqueness assertion is clear. □

If R is a $\mathbf{Z}$-torsion-free ring, then there is an injective ring homomorphism $\phi \colon R \to R \otimes \mathbf{Q}$ defined by $\phi(r) = r \otimes 1$ for $r \in R$. Since $R \otimes \mathbf{Q}$ is a $\mathbf{Q}$-algebra, the above Proposition leads to the following result.

Corollary 4.3. *Let R be a $\mathbf{Z}$-torsion-free ring. If the elements $\mathbf{a} +_W \mathbf{b}$ and $\mathbf{a} \times_W \mathbf{b}$ exist in $\mathbf{W}(R)$ such that the two conditions in* (4.2) *are satisfied, then they must be unique.*

The following technical result due to Dwork [Dwork (1958)] will imply that the Witt operations $+_W$ and $\times_W$ exist in any ring, not just $\mathbf{Q}$-algebras. This result will be used several more times later in this chapter. In [Hazewinkel (1978)] it appears as Lemma (17.6.1), and the proof there is different from the one given below. Dwork's Theorem is proved in [Hazewinkel (1978)] by employing the so-called Functional Equation Lemma, which is about some integrality conditions of power series related to the logarithm of a formal group law. The proof given below is more direct and does not mention formal group laws. It is close to the proof of (17.1.3) in [Hazewinkel (1978)].

Theorem 4.4 (Dwork's Theorem). *Let R be a $\mathbf{Z}$-torsion-free ring. Suppose that for each prime p, there exists a ring endomorphism $\sigma_p \colon R \to R$ such that $\sigma_p(x) \equiv x^p$ (mod pR) for all $x \in R$. Then an element $\mathbf{b} = (b_i) \in \mathbf{W}(R)$ has the form*

$$\mathbf{b} = (w_1(\mathbf{a}), w_2(\mathbf{a}), w_3(\mathbf{a}), \ldots)$$

for some $\mathbf{a} \in \mathbf{W}(R)$ *if and only if*

$$b_{p^n m} \equiv \sigma_p\left(b_{p^{n-1}m}\right) \pmod{p^n R} \tag{4.3}$$

whenever p *is a prime and* m *is relatively prime to* p*. If the element* $\mathbf{a}$ *exists, then it must be unique, and* a_n *depends only on* $b_1, \ldots, b_n$*.*

To prove Dwork's Theorem, we first need some preliminary results.

Lemma 4.5. *Let* R *be a* $\mathbf{Z}$*-torsion-free ring, and let* a *and* b *be elements in* R *such that* $a \equiv b \pmod{pR}$ *for some prime* p*. Then* $a^{p^{k-1}} \equiv b^{p^{k-1}} \pmod{p^k R}$ *for all* $k \geq 1$*.*

Proof. The proof is by induction. The assertion is true for $k = 1$ by assumption. Suppose that $a^{p^i} \equiv b^{p^i} \pmod{p^{i+1}R}$ for $i \leq k-2$. Thus, we have $a^{p^{k-2}} = b^{p^{k-2}} + p^{k-1}r$ for some element $r \in R$. Therefore, raising both sides of this equality to the pth power, we obtain:

$$\begin{aligned} a^{p^{k-1}} &= \left(b^{p^{k-2}} + p^{k-1}r\right)^p \\ &= b^{p^{k-1}} + pb^{(p-1)p^{k-2}}(p^{k-1}r) + \binom{p}{2} b^{(p-2)p^{k-2}}(p^{k-1}r)^2 \\ &\quad + \cdots + (p^{k-1}r)^p \\ &\equiv b^{p^{k-1}} \pmod{p^k R}, \end{aligned}$$

as desired. □

Lemma 4.6. *In the polynomial ring* $R = \mathbf{Z}[x_1, x_2, x_3, \ldots]$*, suppose that for each prime* p*,* $\sigma_p \colon R \to R$ *is a ring endomorphism satisfying* $\sigma_p(f) \equiv f^p \pmod{pR}$ *for all* $f \in R$*. Then the Witt polynomials* $w_n = w_n(\mathbf{x})$ (4.1) *in* R *satisfy*

$$w_{p^n m} \equiv \sigma_p\left(w_{p^{n-1}m}\right) \pmod{p^n R} \tag{4.4}$$

whenever p *is a prime and* m *is relatively prime to* p*.*

Proof. In the polynomial

$$w_{p^n m} = \sum_{d \mid p^n m} d x_d^{\frac{p^n m}{d}},$$

the terms in which d has p^n as a factor become 0 modulo p^n. Therefore, we have

$$w_{p^n m} \equiv \sum_{d \mid p^{n-1}m} d x_d^{\frac{p^n m}{d}} \pmod{p^n R}. \tag{4.5}$$

If d is a positive integer that divides $p^{n-1}m$, then $d = p^k h$ for some $k \leq n-1$ and h dividing m. Since $\sigma_p(x_d) \equiv x_d^p \pmod{pR}$ by assumption, it follows from Lemma 4.5 that

$$\sigma_p(x_d)^{p^{n-1-k}} \equiv x_d^{p^{n-k}} \pmod{p^{n-k}R}.$$

Raising both sides of this congruence equality to the power $\frac{m}{h}$, we obtain

$$\sigma_p(x_d)^{\frac{p^{n-1}m}{d}} \equiv x_d^{\frac{p^n m}{d}} \pmod{p^{n-k}R}.$$

Multiplying by d on both sides, this implies that

$$d\sigma_p(x_d)^{\frac{p^{n-1}m}{d}} \equiv dx_d^{\frac{p^n m}{d}} \pmod{p^n R}. \tag{4.6}$$

Therefore, summing (4.6) over the positive integers d that divide $p^{n-1}m$, we have

$$\begin{aligned}\sigma_p\left(w_{p^{n-1}m}\right) &= \sum_{d|p^{n-1}m} d\sigma_p(x_d)^{\frac{p^{n-1}m}{d}} \\ &\equiv \sum_{d|p^{n-1}m} dx_d^{\frac{p^n m}{d}} \pmod{p^n R},\end{aligned}$$

which is exactly (4.5). □

Proof. [Proof of Theorem 4.4] The uniqueness of $\mathbf{a}$, if it exists, is clear from the proof of Proposition 4.2. The "only if" part of the Theorem follows from Lemma 4.6. In fact, since (4.4) is true in the universal case, i.e., in the polynomial ring $\mathbf{Z}[x_1, x_2, x_3, \ldots]$, it is true in any ring. Therefore, it remains to prove the "if" part.

Thus, we assume that (4.3) is true. Since $w_1(x_1) = x_1$, we take $a_1 = b_1$. The rest of the a_i's are constructed by induction. Suppose that $a_1, \ldots, a_{n-1}$ have been found such that $b_i = w_i(a_1, \ldots, a_i)$ for $1 \leq i \leq n-1$. We want to find $a_n \in R$ such that

$$b_n = w_n(a_1, \ldots, a_n) = \chi + na_n,$$

where

$$\chi = \sum_{d|n,\, d<n} da_d^{\frac{n}{d}}.$$

Since χ only depends on $a_1, \ldots, a_{n-1}$, it has already been constructed. Therefore, a_n exists if and only if

$$b_n \equiv \chi \pmod{nR}. \tag{4.7}$$

Suppose that the prime decomposition of n is $n = p_1^{e_1} \cdots p_k^{e_k}$, where $p_1, \ldots, p_k$ are distinct primes and each $e_i \geq 1$. To show (4.7), it suffices to show

$$b_n \equiv \chi \pmod{p_i^{e_i} R} \tag{4.8}$$

for $i = 1, \ldots, k$ because R is $\mathbf{Z}$-torsion-free.

Pick a prime $p \in \{p_1, \ldots, p_k\}$, and write p^e for the highest power of p that divides n. By assumption (4.3), we have

$$b_n \equiv \sigma_p\left(b_{\frac{n}{p}}\right) \pmod{p^e R}.$$

So to prove (4.8), it suffices to show that

$$\chi = \sum_{d|n,\, d<n} da_d^{\frac{n}{d}} \equiv \sigma_p\left(b_{\frac{n}{p}}\right) \pmod{p^e R}. \tag{4.9}$$

In the above sum, if d contains a factor of p^e, then the corresponding summand is 0 modulo p^e. So we have

$$\chi \equiv \sum_{d \mid \frac{n}{p}} d a_d^{\frac{n}{d}} \pmod{p^e R}. \tag{4.10}$$

For the right-hand side of (4.9), note that the induction hypothesis applies to $b_{\frac{n}{p}}$, giving

$$b_{\frac{n}{p}} = \sum_{d \mid \frac{n}{p}} d a_d^{\frac{n}{pd}}.$$

Applying σ_p to both sides, we obtain

$$\sigma_p\left(b_{\frac{n}{p}}\right) = \sum_{d \mid \frac{n}{p}} d \sigma_p(a_d)^{\frac{n}{pd}}. \tag{4.11}$$

In the above sum, for a typical d, write p^r for the highest power of p that divides d. Comparing (4.10) and (4.11), we see that in order to prove (4.9), it suffices to show

$$\sigma_p\left(a_d\right)^{\frac{n}{pd}} \equiv a_d^{\frac{n}{d}} \pmod{p^{e-r} R}.$$

To prove this last condition, it is enough to prove

$$\sigma_p\left(a_d\right)^{p^{e-r-1}} \equiv a_d^{p^{e-r}} \pmod{p^{e-r} R},$$

which follows from the assumption $\sigma_p(a_d) \equiv a_d^p \pmod{pR}$ and Lemma 4.5. This shows that a_n exists and finishes the induction. The assertion that a_n only depends on $b_1, \ldots, b_n$ is evident from the proof above.

The proof of Theorem 4.4 is complete. □

Dwork's Theorem 4.4 allows us to define the Witt operations $+_W$ and $\times_W$ (4.2) in the universal case. Let $\{x_i\}$ and $\{y_i\}$ $(i \geq 1)$ be two countably infinite sets of variables.

Corollary 4.7. *In the polynomial ring*

$$R = \mathbf{Z}[x_1, x_2, x_3, \ldots; y_1, y_2, y_3, \ldots],$$

there exist unique polynomials

$$\begin{aligned} \xi_n &= \xi_n(x_1, \ldots, x_n; y_1, \ldots, y_n), \\ \pi_n &= \pi_n(x_1, \ldots, x_n; y_1, \ldots, y_n), \\ \iota_n &= \iota_n(x_1, \ldots, x_n) \end{aligned} \tag{4.12}$$

for $n \geq 1$ *such that*

$$\begin{aligned} w_n(\xi_1, \ldots, \xi_n) &= w_n(\mathbf{x}) + w_n(\mathbf{y}), \\ w_n(\pi_1, \ldots, \pi_n) &= w_n(\mathbf{x}) \cdot w_n(\mathbf{y}), \\ w_n(\iota_1, \ldots, \iota_n) &= -w_n(\mathbf{x}) \end{aligned} \tag{4.13}$$

for $n \geq 1$.

Proof. If p is a prime, let $\sigma_p\colon R \to R$ be the ring endomorphism defined by $\sigma_p(x_i) = x_i^p$ and $\sigma_p(y_i) = y_i^p$ for $i \geq 1$. Since the pth power map is linear modulo p, it is clear that $\sigma_p(f) \equiv f^p \pmod{pR}$ for all $f \in R$. Therefore, we can apply Dwork's Theorem 4.4.

First we apply Dwork's Theorem with

$$\mathbf{b} = (w_1(\mathbf{x}) + w_1(\mathbf{y}), w_2(\mathbf{x}) + w_2(\mathbf{y}), w_3(\mathbf{x}) + w_3(\mathbf{y}), \ldots).$$

We need to check (4.3) when p and m are relatively prime, which means

$$w_{p^n m}(\mathbf{x}) + w_{p^n m}(\mathbf{y}) \equiv \sigma_p\left(w_{p^{n-1}m}(\mathbf{x})\right) + \sigma_p\left(w_{p^{n-1}m}(\mathbf{y})\right) \pmod{p^n R}.$$

It suffices to show that

$$w_{p^n m}(\mathbf{x}) \equiv \sigma_p\left(w_{p^{n-1}m}(\mathbf{x})\right) \pmod{p^n R}. \tag{4.14}$$

This congruence is true by Lemma 4.6. This shows the existence and uniqueness of the ξ_n's.

For π_n and ι_n, we apply Dwork's Theorem 4.4 to

$$\mathbf{b} = (w_1(\mathbf{x}) \cdot w_1(\mathbf{y}), w_2(\mathbf{x}) \cdot w_2(\mathbf{y}), w_3(\mathbf{x}) \cdot w_3(\mathbf{y}), \ldots)$$

and

$$\mathbf{b} = (-w_1(\mathbf{x}), -w_2(\mathbf{x}), -w_3(\mathbf{x}), \ldots),$$

respectively. We need to check (4.3), which in these two cases mean:

$$w_{p^n m}(\mathbf{x}) \cdot w_{p^n m}(\mathbf{y}) \equiv \sigma_p\left(w_{p^{n-1}m}(\mathbf{x}) \cdot w_{p^{n-1}m}(\mathbf{y})\right) \pmod{p^n R},$$
$$-w_{p^n m}(\mathbf{x}) \equiv \sigma_p\left(-w_{p^{n-1}m}(\mathbf{y})\right) \pmod{p^n R}.$$

Both of these conditions follow immediately from (4.14). □

Example 4.8. Since $w_1(\mathbf{x}) = x_1$, it is clear that $\xi_1 = x_1 + y_1$, $\pi_1 = x_1 y_1$, and $\iota_1 = -x_1$. Here we compute ξ_p, π_p, and ι_p when p is a prime. First recall that $w_p(\mathbf{x}) = x_1^p + p x_p$. So the condition

$$w_p(\xi_1, \ldots, \xi_p) = w_p(\mathbf{x}) + w_p(\mathbf{y})$$

is equivalent to

$$(x_1 + y_1)^p + p\xi_p = (x_1^p + p x_p) + (y_1^p + p y_p).$$

This implies that

$$\xi_p = x_p + y_p - \frac{1}{p} \sum_{i=1}^{p-1} \binom{p}{i} x_1^{p-i} y_1^i.$$

Similarly, the condition

$$w_p(\pi_1, \ldots, \pi_p) = w_p(\mathbf{x}) w_p(\mathbf{y})$$

means that

$$\begin{aligned}\pi_1^p + p\pi_p &= (x_1^p + p x_p)(y_1^p + p y_p)\\ &= x_1^p y_1^p + p x_p y_1^p + p x_1^p y_p + p^2 x_p y_p.\end{aligned}$$

This gives

$$\pi_p = x_1^p y_p + x_p y_1^p + p x_p y_p.$$

A similar computation gives

$$\iota_p = \begin{cases} -x_1^2 - x_2 & \text{if } p = 2, \\ -x_p & \text{if } p > 2. \end{cases}$$

Note that ξ_p, π_p, and ι_p depend only on x_1, x_p, y_1, and y_p.

One can continue and compute ξ_n, π_n, and ι_n when n is the product of two primes (equal or distinct), then three primes, and so forth. For example, since $w_{p^2}(\mathbf{x}) = x_1^{p^2} + p x_p^p + p^2 x_{p^2}$, the condition

$$w_{p^2}(\iota_1, \ldots, \iota_{p^2}) = -w_{p^2}(\mathbf{x})$$

becomes

$$\iota_1^{p^2} + p\iota_p^p + p^2 \iota_{p^2} = -x_1^{p^2} - p x_p^p - p^2 x_{p^2}.$$

Both ι_1 and ι_p are known, so we can solve for ι_{p^2}:

$$\iota_{p^2} = \begin{cases} -x_1^4 - x_2^2 - x_1^2 x_2 - x_4 & \text{if } p = 2, \\ -x_{p^2} & \text{if } p > 2. \end{cases}$$

The same procedure can be applied to ξ_{p^2} and π_{p^2}, and the computation is similar when n is the product of two distinct primes. This gives an inductive algorithm to compute the polynomials ξ_n, π_n, and ι_n.

Using the universal polynomials ξ_n and π_n in Corollary 4.7, we can now define the Witt addition and multiplication for an arbitrary ring. Recall that $\mathbf{W}(R) = \prod_{i=1}^{\infty} R$ as a set.

Definition 4.9. Let R be a ring. Define the operations

$$+_W, \times_W : \mathbf{W}(R) \times \mathbf{W}(R) \to \mathbf{W}(R)$$

by

$$\begin{aligned} \mathbf{a} +_W \mathbf{b} &= (\xi_1(a_1; b_1), \xi_2(a_1, a_2; b_1, b_2), \ldots), \\ \mathbf{a} \times_W \mathbf{b} &= (\pi_1(a_1; b_1), \pi_2(a_1, a_2; b_1, b_2), \ldots), \end{aligned}$$

where the polynomials ξ_n and π_n (4.12) are the ones in Corollary 4.7.

Theorem 4.10. *Let R be a ring. Then $\mathbf{W}(R)$ is a ring with:*

- *addition* $+_W$,
- *multiplication* $\times_W$,
- *additive zero* $(0, 0, \ldots)$, *and*
- *multiplicative identity* $(1, 0, 0, \ldots)$.

The additive inverse of an element $\mathbf{a} = (a_i) \in \mathbf{W}(R)$ *is given by*

$$-\mathbf{a} = (\iota_1(a_1), \iota_2(a_1, a_2), \ldots),$$

where the polynomials ι_n (4.12) *are as in Corollary 4.7. Moreover, if* $\phi\colon R \to S$ *is a ring homomorphism, then the map* $\mathbf{W}(\phi)\colon \mathbf{W}(R) \to \mathbf{W}(S)$ *defined by*

$$\mathbf{W}(\phi)(a_1, a_2, \ldots) = (\phi(a_1), \phi(a_2), \ldots) \tag{4.15}$$

is a ring homomorphism.

Proof. Recall the Cartesian product ring $R^{\mathbf{N}} = \prod_{i=1}^{\infty} R$ with entry-wise addition and multiplication (3.20). To prove the Theorem, first assume that R is a $\mathbf{Q}$-algebra. In this case, it follows from the proof of Proposition 4.2 that the map $w\colon \mathbf{W}(R) \to R^{\mathbf{N}}$ defined by

$$w(\mathbf{a}) = (w_1(\mathbf{a}), w_2(\mathbf{a}), \ldots)$$

is a bijection. Moreover, this map w preserves addition and multiplication because the conditions (4.13) hold in the universal case. It follows that $\mathbf{W}(R)$ is a ring as well. The assertions concerning the additive zero, multiplicative identity, and additive inverse follow from the identities:

$$\begin{aligned} w(0,0,0,\ldots) &= (0,0,0,\ldots), \\ w(1,0,0,\ldots) &= (1,1,1,\ldots), \\ w(\iota_1(a_1), \iota_2(a_1,a_2), \ldots) &= (-w_1(a_1), -w_2(a_1,a_2), \ldots). \end{aligned}$$

Next suppose that R is a $\mathbf{Z}$-torsion-free ring. Then there is a natural injective ring homomorphism $\phi\colon R \to R \otimes \mathbf{Q}$ from R into a $\mathbf{Q}$-algebra, sending r to $r \otimes 1$ for $r \in R$. The map $\mathbf{W}(\phi)\colon \mathbf{W}(R) \to \mathbf{W}(R \otimes \mathbf{Q})$ is injective and preserves $(0,0,\ldots)$ and $(1,0,0,\ldots)$. Moreover, $\mathbf{W}(\phi)$ preserves $+_W$, $\times_W$, and additive inverse, since they are defined by the universal polynomials ξ_n, π_n, and ι_n with integer coefficients. Since $\mathbf{W}(R \otimes \mathbf{Q})$ is a ring, it follows that $\mathbf{W}(R)$ is a ring as well.

For a general ring R, let S be any $\mathbf{Z}$-torsion-free ring that has R as a homomorphic image. In other words, there is a surjective ring homomorphism $\varphi\colon S \to R$. (The reader should check that such a ring S always exists.) From the previous paragraph, $\mathbf{W}(S)$ is a ring. Moreover, the map $\mathbf{W}(\varphi)\colon \mathbf{W}(S) \to \mathbf{W}(R)$ is surjective and preserves $(0,0,\ldots)$, $(1,0,0,\ldots)$, $+_W$, $\times_W$, and additive inverse. Therefore, $\mathbf{W}(R)$ is a ring, as desired.

The last assertion, that $\mathbf{W}(\phi)$ is a ring homomorphism if ϕ is, is clear from the definition. □

Definition 4.11. The ring $\mathbf{W}(R)$ is called the ring of *big Witt vectors on R*.

Corollary 4.12. *The big Witt vectors construction gives a functor*

$$\mathbf{W}\colon \mathbf{Ring} \to \mathbf{Ring}$$

on the category of rings. Moreover, for each $n \geq 1$, *the Witt polynomial* w_n (4.1) *gives a natural transformation* $w_n\colon \mathbf{W} \to Id$ *of functors on the category of rings.*

Proof. It is immediate from the definition (4.15) that $\mathbf{W}$ is a functor on the category of rings. For each ring R, the map $w_n \colon \mathbf{W}(R) \to R$ defined as

$$w_n(\mathbf{a}) = \sum_{d|n} d a_d^{\frac{n}{d}}$$

is a ring homomorphism because the identities (4.13) hold in the universal case. For a ring homomorphism $\phi \colon R \to S$, the commutativity of the diagram

$$\begin{array}{ccc} \mathbf{W}(R) & \xrightarrow{w_n} & R \\ {\scriptstyle \mathbf{W}(\phi)}\downarrow & & \downarrow{\scriptstyle \phi} \\ \mathbf{W}(S) & \xrightarrow{w_n} & S \end{array}$$

follows immediately from the additivity and multiplicativity of ϕ. This shows that w_n gives a natural transformation. □

The following characterization of the functor $\mathbf{W}$ is an immediate consequence of the proof of Theorem 4.10.

Corollary 4.13. *The big Witt vectors functor* $\mathbf{W} \colon \mathbf{Ring} \to \mathbf{Ring}$ *is uniquely characterized by the following three properties:*

(1) For a ring R, $\mathbf{W}(R) = \prod_{i=1}^{\infty} R$ as a set.
(2) For a ring homomorphism $\phi \colon R \to S$, the map $\mathbf{W}(\phi) \colon \mathbf{W}(R) \to \mathbf{W}(S)$ is defined as ϕ in each coordinate, as in (4.15).
(3) The Witt polynomials $w_n \colon \mathbf{W}(R) \to R$ are ring homomorphisms for $n \geq 1$.

4.2 Artin-Hasse Exponential

Now we want to show that, for any ring R, the universal λ-ring $\Lambda(R)$ is isomorphic to the ring $\mathbf{W}(R)$ of big Witt vectors on R. This is actually true functorially, in the sense that there is a natural isomorphism $E \colon \Lambda \to \mathbf{W}$ of functors on the category of rings. This natural isomorphism is called the *Artin-Hasse Exponential* (Theorem 4.16). As a result, the ring $\mathbf{W}(R)$ has a canonical λ-ring structure (Corollary 4.18).

We first need the following preliminary result.

Lemma 4.14. *Let R be a ring. Then every strict power series $f = 1 + \sum_{i=1}^{\infty} r_i t^i \in \Lambda(R)$ over R can be written uniquely as*

$$f = \prod_{n=1}^{\infty} (1 - (-1)^n a_n t^n). \tag{4.16}$$

Proof. It is clear that any product as in (4.16) is a strict power series over R. On the other hand, expanding the product we obtain:

$$\begin{aligned}\prod_{n=1}^{\infty}(1-(-1)^n a_n t^n) &= (1+a_1 t)(1-a_2 t^2)(1+a_3 t^3)\cdots \\ &= 1 + a_1 t - a_2 t^2 + (-a_1 a_2 + a_3)t^3 + \cdots \\ &= 1 + \sum_{n\geq 1}\Bigg(\sum_{\substack{1\leq i_1<\cdots<i_k\leq n \\ i_1+\cdots+i_k=n}} (-1)^{k+n} a_{i_1}\cdots a_{i_k}\Bigg)t^n.\end{aligned}$$

The sign $(-1)^{k+n}$ comes from the observation that each a_i is attached to the sign $(-1)^{i+1}$. So the sign attached to the product $a_{i_1}\cdots a_{i_k}$ is

$$(-1)^{(i_1+1)+\cdots+(i_k+1)} = (-1)^{k+i_1+\cdots+i_k} = (-1)^{k+n}.$$

Therefore, given f as in the Lemma, a_n is uniquely determined by $r_1, \ldots, r_n$. Indeed, we must have $a_1 = r_1$, $a_2 = -r_2$, and $a_3 = r_3 - r_1 r_2$. In general, the coefficient of t^n in the product is

$$(-1)^{n+1}a_n + \sum_{\substack{1\leq i_1<\cdots<i_k<n \\ i_1+\cdots+i_k=n}} (-1)^{k+n} a_{i_1}\cdots a_{i_k}.$$

By induction every term a_{i_j} appearing in the above sum is uniquely determined by r_i for $i \leq i_j < n$. Therefore, we can write a_n in terms of $r_1, \ldots, r_n$ as

$$a_n = (-1)^{n+1}r_n + \sum_{\substack{1\leq i_1<\cdots<i_k<n \\ i_1+\cdots+i_k=n}} (-1)^{k} a_{i_1}\cdots a_{i_k}$$

So every strict power series over R can be written in the form (4.16). The uniqueness part is clear. □

Example 4.15. We have

$$1 - 3t + 4t^2 + 2t^3 + \cdots = (1-3t)(1+4t^2)(1+14t^3)\cdots.$$

We use Lemma 4.14 to define the desired isomorphism from $\Lambda(R)$ to $\mathbf{W}(R)$. Recall from §2.2.3 that a natural isomorphism is a natural transformation that is an isomorphism at every object.

Theorem 4.16 (The Artin-Hasse Exponential). *There is a natural isomorphism $E\colon \Lambda \to \mathbf{W}$ of functors on the category of rings. For a ring R, the ring isomorphism $E_R\colon \Lambda(R) \to \mathbf{W}(R)$ is given by*

$$E_R\Bigg(\prod_{n=1}^{\infty}(1-(-1)^n a_n t^n)\Bigg) = (a_1, a_2, a_3, \ldots).$$

Proof. By Lemma 4.14 the map E_R is a well-defined function and is a bijection. Next we show that E_R is a ring homomorphism, which will imply that it is a ring isomorphism.

First observe that E_R preserves the additive zero and the multiplicative identity, since $E_R(1) = (0, 0, 0, \ldots)$ and $E_R(1 + t) = (1, 0, 0, \ldots)$. Now we show that E_R is additive. Consider $f = \prod(1 - (-1)^n a_n t^n)$ and $g = \prod(1 - (-1)^n b_n t^n)$ in $\Lambda(R)$. Write

$$\mathbf{c} = (c_1, c_2, \ldots) = \mathbf{a} +_W \mathbf{b} \in \mathbf{W}(R),$$

where $\mathbf{a} = (a_1, a_2, \ldots)$ and $\mathbf{b} = (b_1, b_2, \ldots)$. Recall that $+_\Lambda$ (2.4) is multiplication of strict power series. Therefore, E_R is additive if and only if

$$\prod_{n\geq 1}(1 - (-1)^n a_n t^n) \cdot \prod_{n\geq 1}(1 - (-1)^n b_n t^n) = \prod_{n\geq 1}(1 - (-1)^n c_n t^n). \tag{4.17}$$

To prove this, fist note that

$$\begin{aligned} -\log\Big(\prod_{n\geq 1}(1 - a_n(-t)^n)\Big) &= -\sum_{n\geq 1}\log(1 - a_n(-t)^n) \\ &= \sum_{n,i\geq 1}(-1)^{ni} n a_n^i \cdot \frac{t^{ni}}{ni} \\ &= \sum_{k\geq 1}\Big(\sum_{n|k}(-1)^k n a_n^{\frac{k}{n}}\Big)\frac{t^k}{k} \\ &= \sum_{k\geq 1}(-1)^k w_k(\mathbf{a}) \cdot \frac{t^k}{k}. \end{aligned} \tag{4.18}$$

Therefore, applying $-\log$ to the left-hand side of (4.17), we obtain:

$$\begin{aligned} &-\log\Big(\prod_{n\geq 1}(1 - (-1)^n a_n t^n) \cdot \prod_{n\geq 1}(1 - (-1)^n b_n t^n)\Big) \\ &= -\log\Big(\prod_{n\geq 1}(1 - a_n(-t)^n)\Big) - \log\Big(\prod_{n\geq 1}(1 - (-1)^n b_n t^n)\Big) \\ &= \sum_{k\geq 1}(-1)^k\left(w_k(\mathbf{a}) + w_k(\mathbf{b})\right) \cdot \frac{t^k}{k} \\ &= \sum_{k\geq 1}(-1)^k w_k(\mathbf{c}) \cdot \frac{t^k}{k}. \end{aligned}$$

By (4.18) again, the last expression is equal to $-\log$ of the right-hand side of (4.17). This shows that E_R is additive.

Finally, we need to show that E_R is multiplicative, i.e.,

$$E_R(f \times_\Lambda g) = E_R(f) \times_W E_R(g). \tag{4.19}$$

As in the proofs of Theorems 2.5 and 2.6, it suffices to prove this in the universal case, when the coefficients of f and g are algebraically independent. In particular, we may assume that their coefficients are the elementary symmetric functions of two countably infinite sets of variables. So we write formally

$$f = \prod_{i\geq 1}(1 + x_i t) = \sum_{i}{}_{\Lambda}(1 + x_i t),$$

$$g = \prod_{j\geq 1}(1 + y_j t) = \sum_{j}{}_{\Lambda}(1 + y_j t),$$

where the x_i and y_j are independent variables. Since we already proved that E_R is additive, it suffices to prove (4.19) when f and g are both 1-dimensional elements (Proposition 2.9), $f = 1 + at$ and $g = 1 + bt$. Since $(1 + at) \times_\Lambda (1 + bt) = 1 + abt$, we have

$$\begin{aligned} E_R\left((1 + at) \times_\Lambda (1 + bt)\right) &= (ab, 0, 0, \ldots) \\ &= (a, 0, 0, \ldots) \times_W (b, 0, 0, \ldots) \\ &= E_R(1 + at) \times_W E_R(1 + bt). \end{aligned}$$

This shows that E_R is multiplicative. We have shown that E_R is a ring homomorphism. Since it is bijective, it must be a ring isomorphism.

If $\phi\colon R \to S$ is a ring homomorphism, then the commutativity of the square

$$\begin{array}{ccc} \Lambda(R) & \xrightarrow{E_R} & \mathbf{W}(R) \\ {\scriptstyle \Lambda(\phi)}\downarrow & & \downarrow{\scriptstyle \mathbf{W}(\phi)} \\ \Lambda(S) & \xrightarrow{E_S} & \mathbf{W}(S) \end{array}$$

is immediate from the definitions of $\Lambda(\phi)$ (2.28) and $\mathbf{W}(\phi)$ (4.15). Therefore, $E\colon \Lambda \to \mathbf{W}$ is a natural isomorphism □

Remark 4.17. In the literature the Artin-Hasse Exponential isomorphism E_R is sometimes defined without the sign $(-1)^n$ (e.g., see (17.2.8) in [Hazewinkel (1978)]). This requires using $(1-t)$ (instead of $(1+t)$) as the multiplicative identity in $\Lambda(R)$. Our sign convention follows that of [Knutson (1973)].

Corollary 4.18. *Let R be a ring. Then the ring $\mathbf{W}(R)$ of big Witt vectors on R has a canonical λ-ring structure.*

Proof. The canonical λ-ring structure on $\Lambda(R)$ (Theorem 2.6) can be pulled over to $\mathbf{W}(R)$ via the Artin-Hasse Exponential isomorphism (Theorem 4.16). In other words, we define the λ-operation $\lambda^n\colon \mathbf{W}(R) \to \mathbf{W}(R)$ for $n \geq 1$ by insisting that the square

$$\begin{array}{ccc} \mathbf{W}(R) & \xrightarrow{\lambda^n} & \mathbf{W}(R) \\ {\scriptstyle E_R^{-1}}\downarrow{\scriptstyle \cong} & & {\scriptstyle \cong}\uparrow{\scriptstyle E_R} \\ \Lambda(R) & \xrightarrow[\lambda^n]{} & \Lambda(R) \end{array}$$

be commutative. The λ-ring axioms (Definition 1.10) on $\mathbf{W}(R)$ follow immediately from those on $\Lambda(R)$ because E_R is a ring isomorphism. □

4.3 Characterization of the Functor Λ

In this section we discuss the analogs of the Witt polynomials w_n for the universal λ-ring functor Λ (Theorem 4.21). We will then be able to characterize the functor Λ along the lines of Corollary 4.13 for the functor $\mathbf{W}$ (Corollary 4.22).

4.3.1 *Analogs of the Witt polynomials*

From Corollary 4.12 we know that the big Witt vectors functor $\mathbf{W}$ comes with a sequence of natural transformations $w_n \colon \mathbf{W} \to Id$, where w_n is the nth Witt polynomial (4.1). Since the universal λ-ring functor Λ is naturally isomorphic to $\mathbf{W}$ via the Artin-Hasse Exponential isomorphism E (Theorem 4.16), there must be natural transformations $z_n \colon \Lambda \to Id$ corresponding to the w_n. Here we construct these natural transformations explicitly.

Let R be a ring. For $n \geq 1$ define the functions

$$z_n \colon \Lambda(R) \to R \tag{4.20}$$

by the generating function

$$z_{-t}(f) = -t\frac{d}{dt}(\log f) = \frac{-tf'}{f}, \tag{4.21}$$

where $f \in \Lambda(R)$ and $z_t(f) = \sum_{n\geq 1} z_n(f)t^n$. In other words, $z_n(f)$ is the coefficient of $(-t)^n$ on the right-hand side of (4.21).

Example 4.19. For a strict power series $f = 1 + a_1t + a_2t^2 + a_3t^3 + \cdots \in \Lambda(R)$, its multiplicative inverse (2.3) is

$$\begin{aligned} f^{-1} &= 1 + \sum_{m\geq 1} i_m(a_1, \ldots, a_m)t^m \\ &= 1 - a_1t + (a_1^2 - a_2)t^2 + (-a_1^3 + 2a_1a_2 - a_3)t^3 \cdots . \end{aligned}$$

Thus, we have

$$\begin{aligned} z_{-t}(f) &= \frac{-tf'}{f} \\ &= (-a_1t - 2a_2t^2 - 3a_3t^3 - \cdots)(1 - a_1t + (a_1^2 - a_2)t^2 + \cdots) \\ &= -a_1t + (a_1^2 - 2a_2)t^2 + (-a_1^3 + 3a_1a_2 - 3a_3)t^3 \\ &\quad + (a_1^4 - 4a_1^2a_2 + 2a_2^2 + 4a_1a_3 - 4a_4)t^4 + \cdots . \end{aligned}$$

Taking the coefficients of $(-t)^n$, we have

$$\begin{aligned} z_1(f) &= a_1, \\ z_2(f) &= a_1^2 - 2a_2, \\ z_3(f) &= a_1^3 - 3a_1a_2 + 3a_3, \\ z_4(f) &= a_1^4 - 4a_1^2a_2 + 2a_2^2 + 4a_1a_3 - 4a_4. \end{aligned}$$

Note that if we set $\deg(a_n) = n$, then $z_n(f)$ is a homogeneous polynomial of degree n in $a_1, \ldots, a_n$. In fact, we have

$$\frac{-tf'}{f} = \sum_{n \geq 1} \Big(\sum_{k+l=n} -ka_k \cdot i_l(a_1, \ldots, a_l) \Big) t^n,$$

so

$$z_n(f) = (-1)^{n+1} \sum_{k+l=n} ka_k \cdot i_l(a_1, \ldots, a_l). \tag{4.22}$$

Since i_l (2.3) is homogeneous of degree l, it follows that $z_n(f)$ is homogeneous of degree n.

The definition of z_{-t} looks similar to the definition (3.1) of ψ_{-t} for the Adams operations. Indeed, they are very closely related.

Proposition 4.20. *For any λ-ring R, the diagram*

$$\begin{array}{ccc} R & \xrightarrow{\lambda_t} & \Lambda(R) \\ {\scriptstyle \psi^n} \downarrow & \swarrow {\scriptstyle z_n} & \\ R & & \end{array} \tag{4.23}$$

is commutative.

Proof. For any element $x \in R$, we have

$$\psi_{-t}(x) = -t \frac{d}{dt} (\log(\lambda_t(x))) = z_{-t}(\lambda_t(x)),$$

which proves the Proposition. □

One way to rephrase Proposition 4.20 is that the Adams operation ψ^n factors through the operation z_n. Moreover, for each $n \geq 1$, this can be accomplished by the structure map λ_t of R. Indeed, it is not hard to prove that, when R is $\mathbf{Z}$-torsion-free, λ_t is the *only* map that makes the diagram (4.23) commutative for all $n \geq 1$. We leave the proof of this fact as an exercise for the reader.

The following result says that z_n is a ring homomorphism and corresponds to the Witt polynomial w_n via the Artin-Hasse Exponential isomorphism E.

Theorem 4.21. *For each $n \geq 1$, the function z_n (4.20) defines a natural transformation $z_n \colon \Lambda \to Id$ of functors on the category of rings. Moreover, the following diagram of natural transformations is commutative,*

$$\begin{array}{ccc} \Lambda & \xrightarrow[\cong]{E} & \mathbf{W}\,, \\ {\scriptstyle z_n} \downarrow & \swarrow {\scriptstyle w_n} & \\ Id & & \end{array} \tag{4.24}$$

where E is the Artin-Hasse Exponential isomorphism (Theorem 4.16).

Proof. First observe that if $f = 1+\sum_{n\geq 1} a_n t^n \in \Lambda(R)$, then $z_n(f)$ is a polynomial with integer coefficients in $a_1, \ldots, a_n$. Therefore, once we show that $z_n\colon \Lambda(R) \to R$ is a ring homomorphism, it will follow immediately from the definition of $\Lambda(\phi)$ (2.28) that $z_n\colon \Lambda \to Id$ is a natural transformation.

To show that z_n is a ring homomorphism, first note that $z_{-t}(1) = -t\frac{d}{dt}(\log(1)) = 0$, so $z_n(1) = 0$, and z_n preserves the additive zero. Likewise, since

$$z_{-t}(1+at) = -t\frac{d}{dt}(\log(1+at)) = \frac{-at}{1+at},$$

we have

$$z_t(1+at) = \frac{at}{1-at} = at + a^2t^2 + a^3t^3 + \cdots .$$

This implies that

$$z_n(1+at) = a^n. \tag{4.25}$$

Setting $a = 1$, we have $z_n(1+t) = 1$, and z_n preserves the multiplicative identity.

To show that z_n is additive, observe that, since $+_\Lambda$ is multiplication of strict power series, we have

$$\begin{aligned} z_{-t}(f +_\Lambda g) &= -t\frac{d}{dt}(\log(f \cdot g)) \\ &= -t\frac{d}{dt}(\log(f) + \log(g)) \\ &= z_{-t}(f) + z_{-t}(g). \end{aligned}$$

So each z_n is additive.

Next we show that z_n is multiplicative. As in the proof of Theorem 4.16 for the multiplicativity of E_R, it suffices to prove it for 1-dimensional elements in $\Lambda(R)$ (Proposition 2.9). Since $(1+at) \times_\Lambda (1+bt) = 1 + abt$, it follows from (4.25) that

$$\begin{aligned} z_n((1+at) \times_\Lambda (1+bt)) &= z_n(1+abt) = (ab)^n = a^n b^n \\ &= z_n(1+at) \cdot z_n(1+bt). \end{aligned}$$

We have shown that $z_n\colon \Lambda(R) \to R$ is a ring homomorphism, which implies immediately that $z_n\colon \Lambda \to Id$ is a natural transformation.

Finally, we show the commutativity of the diagram (4.24). Since z_n, E_R, and w_n are all additive, once again it suffices to show $z_n = w_n E_R$ for 1-dimensional elements in $\Lambda(R)$. We have

$$\begin{aligned} w_n E_R(1+at) &= w_n(a, 0, 0, \ldots) = a^n \\ &= z_n(1+at), \end{aligned}$$

as desired. □

Using the characterization of the functor **W** (Corollary 4.13) together with Theorems 4.16 and 4.21, we obtain the following analogous characterization of the functor Λ.

Corollary 4.22. *The universal λ-ring functor Λ:* **Ring** $\to$ **Ring** *is uniquely characterized by the following three properties:*

(1) *For a ring R, $\Lambda(R)$ is the set of all strict power series over R.*
(2) *For a ring homomorphism $\phi\colon R \to S$, the map $\Lambda(\phi)\colon \Lambda(R) \to \Lambda(S)$ is defined as*

$$\Lambda(\phi)(1 + a_1t + a_2t^2 + \cdots) = 1 + \phi(a_1)t + \phi(a_2)t^2 + \cdots .$$

(3) *The functions $z_n\colon \Lambda(R) \to R$ defined in* (4.20) *are ring homomorphisms for $n \geq 1$.*

4.4 W as a Comonad

The comonad structure on the universal λ-ring functor Λ (Theorem 2.25) can be pulled over to the big Witt vectors functor $\mathbf{W}$ using the Artin-Hasse Exponential isomorphism E (Theorem 4.16). Indeed, by structure transport, the comultiplication $\Delta\colon \mathbf{W} \to \mathbf{W}^2$ on $\mathbf{W}$ is defined by the commutative diagram

$$\begin{array}{ccccc} \mathbf{W}(R) & & \xrightarrow{\quad\Delta_R\quad} & & \mathbf{W}^2(R) \\ {\scriptstyle E_R^{-1}}\big\downarrow{\scriptstyle \cong} & & & & {\scriptstyle \cong}\big\uparrow{\scriptstyle E_{\mathbf{W}(R)}} \\ \Lambda(R) & \xrightarrow[\lambda_t]{} & \Lambda^2(R) & \xrightarrow[\Lambda(E_R)]{\cong} & \Lambda(\mathbf{W}(R)) \end{array} \tag{4.26}$$

for a given ring R. In other words, Δ is the composition $\Delta = E^2\lambda_tE^{-1}$. The counit on $\mathbf{W}$ is given by the first Witt polynomial $z_1E^{-1} = w_1\colon \mathbf{W} \to Id$, and z_1 is exactly the counit ε on Λ. The comonad axioms (Definition 2.19) for $(\mathbf{W}, \Delta, w_1)$ follow immediately from those for $(\Lambda, \lambda, \varepsilon)$, the universal λ-ring comonad.

The purpose of this section is to give a more explicit description of the comultiplication Δ. As we will see below, Δ is uniquely characterized by the Witt polynomial w_n and a certain natural transformation $\mathbf{f}_n\colon \mathbf{W} \to \mathbf{W}$, called the *Frobenius map*, for $n \geq 1$. The Frobenius map $\mathbf{f}_n$ corresponds, via the Artin-Hasse Exponential isomorphism, to the Adams operation ψ^n on the universal λ-ring.

We first construct the Frobenius natural transformation $\mathbf{f}_n^\Lambda\colon \Lambda \to \Lambda$ for the universal λ-ring functor (Theorem 4.26). Then we identify it with the Adams operation ψ^n (Theorem 4.27). This $\mathbf{f}_n^\Lambda$ is pulled over to $\mathbf{f}_n$ on $\mathbf{W}$ using the Artin-Hasse Exponential isomorphism E (Definition 4.28). Then we define the comultiplication Δ on $\mathbf{W}$ (Theorem 4.30) using $\mathbf{f}_n$ and Dwork's Theorem 4.4.

Most of the materials in this section can also be found in [Hazewinkel (1978)]. However, we present the materials differently. In particular, a crucial result for defining Δ explicitly is the property (4.34), which states that $\mathbf{f}_p$ is the pth power map modulo $p\mathbf{W}(R)$. This is proved in (17.6.10) in [Hazewinkel (1978)] by a direct computation together with Dwork's Theorem 4.4. On the other hand, our proof of this property is very simple, since we can make use of the corresponding result for the Adams operation ψ^p (3.4) on the universal λ-ring $\Lambda(R)$. At the end of this section we will give an alternative proof of (4.34) following [Hazewinkel (1978)].

4.4.1 *The polynomials* $Q_{n,m}$

In order to define the Frobenius maps for $\Lambda(R)$, we first need certain universal polynomials $Q_{n,m}$, which are defined as follows.

Let $x_1, x_2, \ldots$ be an infinite sequence of variables, and let $s_1, s_2, \ldots$ be their elementary symmetric functions. Recall from (1.12) that this means that

$$s_n = \sum_{i_1 < \cdots < i_n} x_{i_1} \cdots x_{i_n}.$$

Thus, we have

$$\begin{aligned} s_1 &= x_1 + x_2 + x_3 + \cdots, \\ s_2 &= \sum_{i<j} x_i x_j, \\ s_3 &= \sum_{i<j<k} x_i x_j x_k, \end{aligned}$$

and so forth.

For a given integer $n \geq 1$, consider the symmetric function

$$\prod_{i=1}^{\infty}(1 + x_i^n t) = 1 + \Big(\sum_{i\geq 1} x_i^n\Big)t + \Big(\sum_{i<j} x_i^n x_j^n\Big)t^2 + \cdots$$

in $x_1, x_2, \ldots$. By the Fundamental Theorem of Symmetric Functions, there exist unique polynomials $Q_{n,m}$ ($m \geq 1$) with integer coefficients such that

$$\prod_{i=1}^{\infty}(1 + x_i^n t) = 1 + \sum_{m\geq 1} Q_{n,m}(s_1, s_2, \ldots)t^m. \tag{4.27}$$

Example 4.23. If $n = 1$, then $\prod(1 + x_i t)$ is the generating function of the elementary symmetric functions. Therefore, we have

$$Q_{1,m}(s_1, s_2, \ldots) = s_m$$

for $m \geq 1$. Likewise, since

$$\sum_{i\geq 1} x_i^2 = \Big(\sum_{i\geq 1} x_i\Big)^2 - 2\sum_{i<j} x_i x_j,$$

we have

$$Q_{2,1}(s_1, s_2, \ldots) = s_1^2 - 2s_2.$$

In general, since

$$Q_{n,m}(s_1, s_2, \ldots) = \sum_{i_1 < \cdots < i_m} x_{i_1}^n \cdots x_{i_m}^n$$

is homogeneous of degree nm and since s_j is homogeneous of degree j, the polynomial $Q_{n,m}(s_1, s_2, \ldots)$ involves only $s_1, \ldots, s_{nm}$. Also note that $Q_{n,1}$ is equal to the polynomial Q_n (3.5) in Theorem 3.9.

4.4.2 *Frobenius maps on* $\Lambda(R)$

We can now define the Frobenius maps on the universal λ-ring.

Definition 4.24. Let R be a ring, and let $n \geq 1$ be an integer. Define a function $\mathbf{f}_n^\Lambda \colon \Lambda(R) \to \Lambda(R)$ by setting

$$\mathbf{f}_n^\Lambda(1 + a_1 t + a_2 t^2 + \cdots) = 1 + \sum_{m \geq 1} Q_{n,m}(a_1, a_2, \ldots) t^m,$$

where $Q_{n,m}$ is the universal polynomial defined in (4.27). The map $\mathbf{f}_n^\Lambda$ is called the *nth Frobenius map* on $\Lambda(R)$.

Example 4.25. In the universal case, we can take a_n as the nth elementary symmetric function on the variables x_i ($i \geq 1$), i.e., $a_n = s_n$. Then we have

$$1 + a_1 t + a_2 t^2 + \cdots = \prod_{i=1}^{\infty} (1 + x_i t),$$

and

$$\begin{aligned} \mathbf{f}_n^\Lambda\left(\prod_{i=1}^{\infty}(1 + x_i t)\right) &= 1 + \sum_{m \geq 1} Q_{n,m}(a_1, a_2, \ldots) t^m \\ &= \prod_{i=1}^{\infty}(1 + x_i^n t) \end{aligned} \tag{4.28}$$

by the definition (4.27) of $Q_{n,m}$.

Theorem 4.26. *For each $n \geq 1$, the nth Frobenius map defines a natural transformation $\mathbf{f}_n^\Lambda \colon \Lambda \to \Lambda$ on the universal λ-ring functor Λ.*

Proof. Since $\mathbf{f}_n^\Lambda$ is defined by some universal polynomials, once we show that it is a ring homomorphism when applied to an arbitrary ring R, it will follow immediately that $\mathbf{f}_n^\Lambda$ is a natural transformation.

Let R be a ring. To show that $\mathbf{f}_n^\Lambda \colon \Lambda(R) \to \Lambda(R)$ is a ring homomorphism, we proceed as in the proof of Theorem 4.21. First note that $\mathbf{f}_n^\Lambda(1) = \mathbf{f}_n^\Lambda\left(\prod(1 + 0t)\right) = 1$ and $\mathbf{f}_n^\Lambda(1 + t) = 1 + t$ by (4.28). Thus, $\mathbf{f}_n^\Lambda$ preserves the additive zero and the multiplicative identity.

Next we show that $\mathbf{f}_n^\Lambda$ is additive. It suffices to consider the universal case, when the coefficients of the two elements in $\Lambda(R)$ under consideration are algebraically independent. In particular, we may assume that they are the elementary symmetric functions of two sets of variables, i.e.,

$$\begin{aligned} 1 + a_1 t + a_2 t^2 + \cdots &= \prod_{i=1}^{\infty}(1 + x_i t), \\ 1 + b_1 t + b_2 t^2 + \cdots &= \prod_{i=1}^{\infty}(1 + y_i t), \end{aligned}$$

where $x_1, x_2, \ldots, y_1, y_2, \ldots$ are independent variables. Recall that the addition $+_\Lambda$ in $\Lambda(R)$ is multiplication of strict power series (2.4). Therefore, we have

$$\prod_{i=1}^{\infty}(1+x_it) +_\Lambda \prod_{i=1}^{\infty}(1+y_it) = \prod_{i=1}^{\infty}(1+x_it)(1+y_it).$$

Using (4.28) we have

$$\begin{aligned}
\mathbf{f}_n^\Lambda\left(\prod_{i=1}^{\infty}(1+x_it)(1+y_it)\right) &= \prod_{i=1}^{\infty}(1+x_i^nt)(1+y_i^nt) \\
&= \prod_{i=1}^{\infty}(1+x_i^nt) +_\Lambda \prod_{i=1}^{\infty}(1+y_i^nt) \\
&= \mathbf{f}_n^\Lambda\left(\prod_{i=1}^{\infty}(1+x_it)\right) +_\Lambda \mathbf{f}_n^\Lambda\left(\prod_{i=1}^{\infty}(1+y_it)\right).
\end{aligned}$$

This shows that $\mathbf{f}_n^\Lambda$ is additive.

Next we show that $\mathbf{f}_n^\Lambda$ is multiplicative. Since we already know that $\mathbf{f}_n^\Lambda$ is additive, it suffices to show the multiplicativity of $\mathbf{f}_n^\Lambda$ for 1-dimensional elements, i.e., linear polynomials in $\Lambda(R)$. In this case, we have

$$\begin{aligned}
\mathbf{f}_n^\Lambda\left((1+at)\times_\Lambda(1+bt)\right) &= \mathbf{f}_n^\Lambda(1+abt) \quad \text{by} \quad (2.5) \\
&= 1+(ab)^nt \quad \text{by} \quad (4.28) \\
&= (1+a^nt)\times_\Lambda(1+b^nt) \\
&= \mathbf{f}_n^\Lambda(1+at)\times_\Lambda \mathbf{f}_n^\Lambda(1+bt).
\end{aligned}$$

This shows that $\mathbf{f}_n^\Lambda$ is multiplicative on 1-dimensional elements, and hence on $\Lambda(R)$.

We have shown that $\mathbf{f}_n^\Lambda$ is a ring homomorphism. By the remarks at the beginning of this proof, this finishes the proof of the Theorem. □

Next we show that the Frobenius natural transformation $\mathbf{f}_n^\Lambda$ is closely related to z_n (4.20) and the Adams operation ψ^n.

Theorem 4.27.

(1) The diagram of natural transformations

$$\begin{array}{ccc}
\Lambda & \xrightarrow{\mathbf{f}_n^\Lambda} & \Lambda \\
{\scriptstyle z_{nm}}\downarrow & \swarrow {\scriptstyle z_m} & \\
Id & &
\end{array} \tag{4.29}$$

is commutative for $n, m \geq 1$.

(2) For any ring R and $n \geq 1$, we have

$$\mathbf{f}_n^\Lambda = \psi^n : \Lambda(R) \to \Lambda(R). \tag{4.30}$$

In particular, the Adams operation ψ^n gives a natural transformation $\psi^n : \Lambda \to \Lambda$ on the universal λ-ring functor Λ.

(3) *For any ring R and prime p, we have*

$$\mathbf{f}_p^\Lambda(g) \equiv \underbrace{g \times_\Lambda \cdots \times_\Lambda g}_{p \text{ factors}} \pmod{p\Lambda(R)} \tag{4.31}$$

for $g \in \Lambda(R)$.

(4) *We have*

$$\mathbf{f}_n^\Lambda \mathbf{f}_m^\Lambda = \mathbf{f}_{nm}^\Lambda \tag{4.32}$$

for $n, m \geq 1$.

Proof. Since $\mathbf{f}_n^\Lambda$, z_m, and z_{nm} are all ring homomorphisms when applied to a ring R, it suffices to prove (4.29) on 1-dimensional elements, i.e., $1 + at \in \Lambda(R)$. In this case, we have

$$\begin{aligned} z_m \mathbf{f}_n^\Lambda(1 + at) &= z_m(1 + a^n t) \quad \text{by} \quad (4.28) \\ &= (a^n)^m \quad \text{by} \quad (4.25) \\ &= z_{nm}(1 + at). \end{aligned}$$

This proves (4.29).

To prove (4.30), again it suffices to show that they agree on 1-dimensional elements because both $\mathbf{f}_n^\Lambda$ and ψ^n are ring homomorphisms. By Proposition 3.5 we know that ψ^n is the nth power map on 1-dimensional elements. On the other hand, we have

$$\begin{aligned} \mathbf{f}_n^\Lambda(1 + at) &= 1 + a^n t \\ &= \underbrace{(1 + at) \times_\Lambda \cdots \times_\Lambda (1 + at)}_{n \text{ factors}} \\ &= \psi^n(1 + at). \end{aligned}$$

Thus, $\mathbf{f}_n^\Lambda$ is also the nth power map on 1-dimensional elements. This proves (4.30).

The assertion (4.31) follows from (4.30) and (3.4), which states that, modulo p, ψ^p is the pth power map.

The assertion (4.32) follows from (4.30) and (3.2), which states that $\psi^n \psi^m = \psi^{nm}$. This finishes the proof of the Theorem. □

4.4.3 *Frobenius maps on* $\mathbf{W}(R)$

Now we define the Frobenius maps on the ring $\mathbf{W}(R)$ of big Witt vectors on R using $\mathbf{f}_n^\Lambda$ and the Artin-Hasse Exponential isomorphism E (Theorem 4.16).

Definition 4.28. Let R be a ring, and let $n \geq 1$ be an integer. Define a function $\mathbf{f}_n \colon \mathbf{W}(R) \to \mathbf{W}(R)$ by insisting that the diagram

$$\begin{array}{ccc} \mathbf{W}(R) & \xrightarrow{\mathbf{f}_n} & \mathbf{W}(R) \\ {\scriptstyle E_R^{-1}} \downarrow \cong & & \cong \uparrow {\scriptstyle E_R} \\ \Lambda(R) & \xrightarrow[\mathbf{f}_n^\Lambda]{} & \Lambda(R) \end{array}$$

be commutative, i.e.,

$$\mathbf{f}_n = E_R \mathbf{f}_n^{\Lambda} E_R^{-1}. \tag{4.33}$$

The function $\mathbf{f}_n$ is called the *nth Frobenius map* on $\mathbf{W}(R)$.

In other words, the Frobenius map $\mathbf{f}_n$ is obtained from $\mathbf{f}_n^{\Lambda}$ by structure transport via the Artin-Hasse Exponential isomorphism E.

Corollary 4.29.

(1) The nth Frobenius map gives a natural transformation $\mathbf{f}_n \colon \mathbf{W} \to \mathbf{W}$ on the big Witt vectors functor $\mathbf{W}$.

(2) The diagram of natural transformations

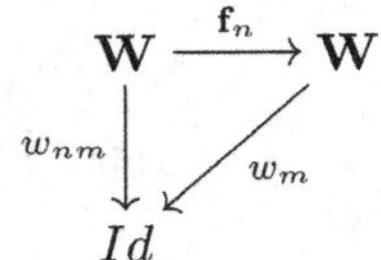

is commutative for $n, m \geq 1$.

(3) For any ring R and prime p, we have

$$\mathbf{f}_p(\mathbf{a}) \equiv \mathbf{a}^p \pmod{p\mathbf{W}(R)} \tag{4.34}$$

for $\mathbf{a} \in \mathbf{W}(R)$.

(4) We have

$$\mathbf{f}_n \mathbf{f}_m = \mathbf{f}_{nm} \tag{4.35}$$

for $n, m \geq 1$.

Proof. All four assertions follow from Theorems 4.21, 4.26, and 4.27. □

In Proposition 4.34 below we will give another proof of (4.34) that does not make use of the Adams operations in the universal λ-ring.

It follows from Corollary 4.29 that, given a ring R, the ring $\mathbf{W}(R)$ of big Witt vectors on R is a ψ-ring (Definition 3.44) when equipped with the Frobenius ring endomorphisms $\mathbf{f}_n \colon \mathbf{W}(R) \to \mathbf{W}(R)$ for $n \geq 1$.

Using the above properties of the Frobenius maps, we can now describe the comultiplication on the functor $\mathbf{W}$.

Theorem 4.30. *There exists a unique natural transformation $\Delta \colon \mathbf{W} \to \mathbf{W}^2$ of functors on the category of rings such that the diagram of natural transformations*

$$\begin{array}{ccc} \mathbf{W} & \xrightarrow{\Delta} & \mathbf{W}^2 \\ {\scriptstyle \mathbf{f}_n}\downarrow & \swarrow {\scriptstyle w_n} & \\ \mathbf{W} & & \end{array} \tag{4.36}$$

is commutative for $n \geq 1$.

In other words, Δ is the unique natural transformation that factors the Frobenius map $\mathbf{f}_n$ through the Witt polynomial $w_n \colon \mathbf{W}^2 \to \mathbf{W}$ for all $n \geq 1$.

The proof of this Theorem will be given after the following preliminary result.

Lemma 4.31. *Let R be a $\mathbf{Z}$-torsion-free ψ-ring (Definition 3.44) such that $\psi^p(a) \equiv a^p \pmod{pR}$ for all elements $a \in R$ and primes p. Then there exists a unique ring homomorphism $D_R \colon R \to \mathbf{W}(R)$ such that*

$$w_n D_R = \psi^n \tag{4.37}$$

for $n \geq 1$.

Proof. Given an element $b \in R$, we set $b_n = \psi^n(b)$. Then for p prime and m relatively prime to p, we have

$$\begin{aligned}\psi^p\left(b_{p^{n-1}m}\right) &= \psi^p(\psi^{p^{n-1}m}(b)) \\ &= \psi^{p^n m}(b) \\ &= b_{p^n m}.\end{aligned}$$

Therefore, Dwork's Theorem 4.4 applies to the element $\mathbf{b} = (b_1, b_2, \ldots) \in \mathbf{W}(R)$, and there exist unique elements $a_1, a_2, \ldots$ in R such that

$$w_n(a_1, \ldots, a_n) = b_n = \psi^n(b)$$

for $n \geq 1$. We now set

$$D_R(b) = (a_1, a_2, \ldots) \in \mathbf{W}(R),$$

which clearly implies that $w_n D_R = \psi^n$ and that D_R is unique. To show that D_R is a ring homomorphism, by the $\mathbf{Z}$-torsion-free assumption on R, it suffices to show that the ψ^n are ring homomorphisms. This is true by hypothesis. □

Proof. [Proof of Theorem 4.30] First we consider the universal case when $R = \mathbf{Z}[x_1, x_2, \ldots]$ is the polynomial ring on an infinite sequence of variables $x_1, x_2, \ldots$. Consider the element $\mathbf{x} = (x_1, x_2, x_3, \ldots) \in \mathbf{W}(R)$. By Corollary 4.29 $\mathbf{W}(R)$ is a $\mathbf{Z}$-torsion-free ψ-ring when equipped with the Frobenius maps $\mathbf{f}_n \colon \mathbf{W}(R) \to \mathbf{W}(R)$ for $n \geq 1$. Moreover, by (4.34) the ψ-ring $\mathbf{W}(R)$ satisfies the hypotheses of Lemma 4.31. Applying (4.37) to the element $\mathbf{x} \in \mathbf{W}(R)$, it follows that there exist unique elements $\Delta_n(\mathbf{x}) \in \mathbf{W}(R)$ for $n \geq 1$ such that

$$\mathbf{f}_n(\mathbf{x}) = w_n\left(\Delta_1(\mathbf{x}), \ldots, \Delta_n(\mathbf{x})\right) \tag{4.38}$$

for all n.

Now for an arbitrary ring A, we define the function $\Delta_A \colon \mathbf{W}(A) \to \mathbf{W}^2(A)$ by setting

$$\Delta_A(\mathbf{a}) = (\Delta_1(\mathbf{a}), \Delta_2(\mathbf{a}), \ldots)$$

for $\mathbf{a} \in \mathbf{W}(A)$. It follows from (4.38) that we have $w_n \Delta_A = \mathbf{f}_n$ for $n \geq 1$. Thus, if A is $\mathbf{Z}$-torsion-free, then Δ_A is the unique ring homomorphism $D_{\mathbf{W}(A)}$ in Lemma 4.31 with $\mathbf{f}_n$ playing the role of ψ^n.

For an arbitrary ring A that is not necessarily $\mathbf{Z}$-torsion-free, we can take a surjective ring homomorphism $\varphi\colon \bar{A} \to A$ in which $\bar{A}$ is $\mathbf{Z}$-torsion-free. Since Δ is defined by universal polynomials, the diagram

$$\begin{array}{ccc} \mathbf{W}(\bar{A}) & \xrightarrow{\Delta_{\bar{A}}} & \mathbf{W}^2(\bar{A}) \\ {\scriptstyle \mathbf{W}(\varphi)} \downarrow & & \downarrow {\scriptstyle \mathbf{W}^2(\varphi)} \\ \mathbf{W}(A) & \xrightarrow{\Delta_A} & \mathbf{W}^2(A) \end{array}$$

is commutative with surjective vertical maps. Since $\Delta_{\bar{A}}$, $\mathbf{W}(\varphi)$, and $\mathbf{W}^2(\varphi)$ are ring homomorphisms, so is Δ_A.

This finishes the proof of Theorem 4.30. □

Corollary 4.32. *The natural transformation Δ in Theorem 4.30 is also given by $\Delta = E^2\lambda_t E^{-1}$ as in (4.26).*

Proof. By Proposition 4.20 and (4.30) we have $z_n\lambda_t = \psi^n = \mathbf{f}_n^\Lambda$ for $n \geq 1$. Applying $E(-)E^{-1}$ to both sides of this equality, we have

$$\begin{aligned} \mathbf{f}_n &= E\mathbf{f}_n^\Lambda E^{-1} \\ &= Ez_n\lambda_t E^{-1} \\ &= w_n E^2\lambda_t E^{-1}. \end{aligned}$$

The Corollary now follows from the uniqueness of Δ. □

Corollary 4.33. *There is a comonad $(\mathbf{W}, \Delta, w_1)$ on the category of rings, where Δ and w_1 are the natural transformations in Theorem 4.30 and Corollary 4.12, respectively.*

Proof. We know that $(\Lambda, \lambda_t, \varepsilon = z_1)$ is a comonad (Theorem 2.25) on the category of rings. By (4.24) and Corollary 4.32, the data (Δ, w_1) is obtained from (λ_t, ε) by structure transport using the Artin-Hasse Exponential isomorphism $E\colon \Lambda \to \mathbf{W}$ from Theorem 4.16. Thus, the comonad axioms for $(\mathbf{W}, \Delta, w_1)$ follow from those for $(\Lambda, \lambda_t, \varepsilon = z_1)$. □

4.4.4 *Alternative proof of* (4.34)

Here we give an alternative, direct proof of (4.34), which states that when p is a prime, the Frobenius map $\mathbf{f}_p\colon \mathbf{W}(R) \to \mathbf{W}(R)$ is the pth power map modulo p on the ring $\mathbf{W}(R)$ of big Witt vectors on R. This is a crucial ingredient in the proof of Theorem 4.30. The proof below is taken from (17.6.10) in [Hazewinkel (1978)].

Proposition 4.34. *For any ring R and prime p, we have*

$$\mathbf{f}_p(\mathbf{a}) \equiv \mathbf{a}^p \pmod{p\mathbf{W}(R)} \tag{4.39}$$

for $\mathbf{a} \in \mathbf{W}(R)$.

Proof. First observe that it suffices to consider the polynomial ring $R = \mathbf{Z}[x_1, x_2, \ldots]$ and the element $\mathbf{x} = (x_1, x_2, \ldots) \in \mathbf{W}(R)$. Indeed, if the Proposition is true in this case, then for a general ring A and an element $\mathbf{a} = (a_i) \in \mathbf{W}(A)$, there exists a unique ring homomorphism $\phi\colon \mathbf{Z}[x_1, x_2, \ldots] \to A$ such that $\phi(x_i) = a_i$ and $\mathbf{W}(\phi)(\mathbf{x}) = \mathbf{a}$. This will imply that the Proposition is true for A as well.

Since R is $\mathbf{Z}$-torsion-free, in order to prove $\mathbf{f}_p(\mathbf{x}) \equiv \mathbf{x}^p \pmod{p\mathbf{W}(R)}$, it suffices to show that

$$w_n\left(\mathbf{f}_p(\mathbf{x}) - \mathbf{x}^p\right) \in pR \tag{4.40}$$

for $n \geq 1$. (The reader should check this.) To prove (4.40) it suffices to show that there exist unique elements $b_i \in R$ $(i \geq 1)$ such that

$$pw_n(b_1, \ldots, b_n) = w_n\left(\mathbf{f}_p(\mathbf{x}) - \mathbf{x}^p\right) \tag{4.41}$$

for $n \geq 1$. To this end, we consider the elements

$$c_n(\mathbf{x}) = p^{-1}w_n\left(\mathbf{f}_p(\mathbf{x}) - \mathbf{x}^p\right) \in \mathbf{Q}[x_1, x_2, \ldots].$$

We claim that they satisfy the following two properties:

$$c_n(\mathbf{x}) \in R = \mathbf{Z}[x_1, x_2, \ldots], \tag{4.42a}$$

$$\sigma_p c_n(\mathbf{x}) \equiv c_{np}(\mathbf{x}) \pmod{p^{\nu_p(n)+1}R} \tag{4.42b}$$

for $n \geq 1$. Here $\sigma_p\colon R \to R$ is the ring endomorphism defined by $\sigma_p(x_i) = x_i^p$ for $i \geq 1$, and $p^{\nu_p(n)}$ is the highest power of p that divides n. If (4.42a) and (4.42b) are true, then by Dwork's Theorem 4.4 there exist unique elements $b_i \in R$ $(i \geq 1)$ such that $w_n(b_1, \ldots, b_n) = c_n(\mathbf{x})$ for $n \geq 1$, which is equivalent to (4.41). Thus, it remains to prove (4.42a) and (4.42b).

For (4.42a) we have

$$\begin{aligned} c_n(\mathbf{x}) &= p^{-1}\left(w_n(\mathbf{f}_p(\mathbf{x})) - w_n(\mathbf{x}^p)\right) \\ &= p^{-1}\left(w_{np}(\mathbf{x}) - w_n(\mathbf{x})^p\right) \end{aligned} \tag{4.43}$$

by Corollary 4.12 and Corollary 4.29 (2). Now we know that

$$w_{np}(\mathbf{x}) \equiv w_n(x_1^p, \ldots, x_n^p) \pmod{p^{\nu_p(n)+1}R} \tag{4.44}$$

by Lemma 4.6. Moreover, we have

$$\begin{aligned} w_n(x_1^p, \ldots, x_n^p) &= \sum_{d|n} d(x_d^p)^{\frac{n}{d}} \\ &\equiv \sum_{d|n} d^p(x_d^p)^{\frac{n}{d}} \pmod{pR} \\ &\equiv w_n(\mathbf{x})^p \pmod{pR}. \end{aligned}$$

Therefore, we have

$$w_{np}(\mathbf{x}) \equiv w_n(\mathbf{x})^p \pmod{pR}.$$

Together with (4.43) this implies the condition (4.42a).

To prove (4.42b), observe that we have

$$\begin{aligned} c_{np}(\mathbf{x}) &= p^{-1}\left(w_{np}(\mathbf{f}_p(\mathbf{x})) - w_{np}(\mathbf{x}^p)\right) \\ &= p^{-1}\left(w_{np^2}(\mathbf{x}) - w_{np}(\mathbf{x})^p\right) \end{aligned} \tag{4.45}$$

as in (4.43). Now by Lemma 4.6 we have

$$w_{np^2}(\mathbf{x}) \equiv w_{np}(x_1^p, \ldots, x_{np}^p) \pmod{p^{\nu_p(n)+2}R}. \tag{4.46}$$

Using Lemma 4.5 on (4.44), we have

$$w_{np}(\mathbf{x})^p \equiv w_n(x_1^p, \ldots, x_n^p)^p \pmod{p^{\nu_p(n)+2}R}. \tag{4.47}$$

Putting (4.46) and (4.47) back into (4.45), we have

$$\begin{aligned} c_{np}(\mathbf{x}) &\equiv p^{-1}\left(w_{np}(x_1^p, \ldots, x_{np}^p) - w_n(x_1^p, \ldots, x_n^p)^p\right) \pmod{p^{\nu_p(n)+1}R} \\ &= c_n(x_1^p, x_2^p, \ldots) \\ &= \sigma_p c_n(\mathbf{x}). \end{aligned}$$

This proves (4.42b).

This finishes the proof of Proposition 4.34. □

The following table provides a dictionary between the functors Λ and $\mathbf{W}$ and their associated natural transformations.

Universal λ-ring	Big Witt vectors	Correspondence
Functor Λ Theorem 2.25	Functor $\mathbf{W}$ Corollary 4.12	Artin-Hasse Exponential $E\colon \Lambda \xrightarrow{\cong} \mathbf{W}$ Theorem 4.16
$z_n\colon \Lambda \to Id$ (4.20)	Witt polynomial $w_n\colon \mathbf{W} \to Id$ (4.1)	$z_n = w_n E$ Theorem 4.21
Frobenius $\mathbf{f}_n^\Lambda\colon \Lambda \to \Lambda$ Theorem 4.26	Frobenius $\mathbf{f}_n\colon \mathbf{W} \to \mathbf{W}$ (4.33)	$\mathbf{f}_n = E\mathbf{f}_n^\Lambda E^{-1}$ (4.33)
Comultiplication $\lambda_t\colon \Lambda \to \Lambda^2$ Theorem 2.25	Comultiplication $\Delta\colon \mathbf{W} \to \mathbf{W}^2$ Theorem 4.30	$\Delta = E^2\lambda_t E^{-1}$ Corollary 4.32

4.5 Exercises

(1) Write down explicitly the polynomials ξ_n and π_n (4.12) when $n = p^2$ or pq, where p and q are distinct primes.

(2) Let $\mathbf{a} = (1, 0, 0, \ldots)$ be the multiplicative identity in $\mathbf{W}(R)$. Compute its additive inverse $\mathbf{b} = -\mathbf{a}$ explicitly.

(3) Let R be a ring. Show that there exists a $\mathbf{Z}$-torsion-free ring S and a surjective ring homomorphism $\phi\colon S \to R$. This was used in the proofs of Theorems 4.10 and 4.30.

(4) Prove the following generalization of Corollary 4.7: Let Θ be any polynomial in two variables. Show that there exist unique polynomials with integer coefficients $\theta_n = \theta_n(x_1, \ldots, x_n; y_1, \ldots, y_n)$ for $n \geq 1$ such that $w_n(\theta_1, \ldots, \theta_n) = \Theta(w_n(\mathbf{x}), w_n(\mathbf{y}))$ for all n.

(5) Give a direct proof of (4.31) without appealing to the Adams operation ψ^p.

(6) Prove (4.32) directly without appealing to the Adams operations on $\Lambda(R)$.

(7) Let R be a $\mathbf{Z}$-torsion-free ring. Show that the map D_R in Lemma 4.31 must be $E\lambda_t$, where $\lambda_t\colon R \to \Lambda(R)$ and E is the Artin-Hasse Exponential isomorphism (Theorem 4.16).

(8) Check that in the proof of Proposition 4.34, it indeed suffices to prove (4.40).

Chapter 5

Binomial Rings

From the point of view of Adams operations, the simplest λ-rings have $\psi^n = Id$ for all n. For example, this is the case for the integers $\mathbf{Z}$ (Example 3.4), the p-local integers $\mathbf{Z}_{(p)}$ (Corollary 3.59), and the p-adic integers $\hat{\mathbf{Z}}_p$ (Corollary 3.63). The purpose of this Chapter is to study λ-rings whose Adams operations are all equal to the identity.

A λ-ring in which all the Adams operations are equal to the identity is equivalent to a *binomial ring*. By definition a binomial ring is a $\mathbf{Z}$-torsion-free ring R such that all the binomial symbols

$$\binom{x}{n} = \frac{x(x-1)\cdots(x-(n-1))}{n!} \in R \otimes \mathbf{Q}$$

actually lie in R for $x \in R$ and $n \geq 1$. For example, the integers $\mathbf{Z}$, the p-local integers $\mathbf{Z}_{(p)}$, and fields of characteristic 0 are binomial rings. When a binomial ring R is considered as a λ-ring, its λ-operations are given by $\lambda^n(x) = \binom{x}{n}$, or equivalently,

$$\lambda_t(x) = (1+t)^x = \sum_{n \geq 0} \binom{x}{n} t^n.$$

The proof that binomial rings are exactly the λ-rings with $\psi^n = Id$ for all n is given in §5.1.

In §5.2 we begin our discussion of integer-valued polynomials. It is shown that every ring $\mathrm{Int}(\mathbf{Z}^X)$ of integer-valued polynomials is a binomial ring (Theorem 5.21). Therefore, every ring of integer-valued polynomials is a λ-ring in which all the Adams operations are equal to the identity map.

In §5.3 a $\mathbf{Z}$-module basis is established for any ring $\mathrm{Int}(\mathbf{Z}^X)$ of integer-valued polynomials (Theorem 5.28). This result is due to [Wilkerson (1982)]. The basis elements consist of the generalized binomial symbols (5.16).

In §5.4 we give another characterization of binomial rings following [Elliott (2006)]. A binomial ring is equivalent to a $\mathbf{Z}$-torsion-free homomorphic image of a ring of integer-valued polynomials (Theorem 5.34). By the end of §5.4 we have the following four equivalent descriptions of the same class of rings (Corollary 5.35):

- λ-rings with $\psi^n = Id$ for all $n \geq 1$.
- Binomial rings.
- $\mathbf{Z}$-torsion-free rings R in which $a^p \equiv a \pmod{pR}$ for $a \in R$ and prime p.
- $\mathbf{Z}$-torsion-free homomorphic images of $\mathrm{Int}(\mathbf{Z}^X)$ for some set X.

In §5.5 we present a result due to [Knutson (1975)], which says that a binomial ring with a certain kind of generating subset must be isomorphic to $\mathbf{Z}$. As immediate applications, Knutson's Theorem implies that if the topological K-theory $K(X)$ of a connected compact Hausdorff space is a binomial ring, then $K(X)$ is isomorphic to $\mathbf{Z}$. Likewise, the only finite group G whose representation ring $R(G)$ is a binomial ring is the trivial group.

In §5.6 we show that, for a binomial ring R, its universal λ-ring $\Lambda(R)$ is isomorphic to the necklace ring $Nr(R)$ of [Metropolis and Rota (1983)]. This result is also due to [Elliott (2006)]. In particular, for a binomial ring R, we have the ring isomorphisms

$$\mathbf{W}(R) \xleftarrow[\cong]{E} \Lambda(R) \xrightarrow[\cong]{E_{Nr}} Nr(R).$$

Here $\mathbf{W}(R)$ is the ring of big Witt vectors on R (Theorem 4.10), E is the Artin-Hasse Exponential isomorphism (Theorem 4.16), and E_{Nr} is the isomorphism to be constructed in §5.6.

5.1 Binomial Rings as λ-Rings

The purpose of this section is to characterize λ-rings in which all the Adams operations are equal to the identity as binomial rings (Theorem 5.3). Another characterization is given in Corollary 5.11. These characterizations are due to [Elliott (2006)], on which this section is based.

At the end of this section, we discuss some consequences of this characterization. In particular, the rings $\mathbf{Z}_{(p)}$ and $\hat{\mathbf{Z}}_p$ of p-local integers and p-adic integers are binomial rings (Corollary 5.12). Moreover, every binomial ring is radical (5.15).

Definition 5.1. A *binomial ring* is a $\mathbf{Z}$-torsion-free ring R such that all the binomial symbols

$$\binom{x}{n} = \frac{x(x-1)\cdots(x-(n-1))}{n!} \in R \otimes \mathbf{Q} \tag{5.1}$$

actually lie in R for $x \in R$ and $n \geq 1$.

In other words, a ring R is binomial if and only if it is $\mathbf{Z}$-torsion-free and is closed under taking binomial symbols.

Example 5.2. Here are some examples of binomial rings.

(1) The ring $\mathbf{Z}$ of integers.

(2) Fields of characteristic 0.
(3) Any $\mathbf{Q}$-algebra.

In the next section, we will see that any ring of integer-valued polynomials is a binomial ring. Moreover, in §5.4 we will show that a binomial ring must be a homomorphic image of some ring of integer-valued polynomials.

Here is the main result of this section.

Theorem 5.3. *A binomial ring has the unique structure of a λ-ring in which all the Adams operations are equal to the identity. Conversely, a λ-ring in which all the Adams operations are equal to the identity is a binomial ring. In either case, the λ-operations are given by*

$$\lambda^n(a) = \binom{a}{n} = \frac{a(a-1)\cdots(a-(n-1))}{n!} \tag{5.2}$$

for $n \geq 1$ and a in the ring.

Since a binomial ring is $\mathbf{Z}$-torsion-free by definition, we have the following immediate consequence.

Corollary 5.4. *A λ-ring in which all the Adams operations are equal to the identity is necessarily $\mathbf{Z}$-torsion-free.*

The proof of Theorem 5.3 involves the following steps:

(1) Show that a binomial ring R satisfies $a^p \equiv a \pmod{pR}$ for $a \in R$ and p prime.
(2) The previous condition implies that a binomial ring R satisfies the hypotheses of Wilkerson's Theorem 3.54 with $\psi^n = Id$ for all n. Thus, the first half of the Theorem follows from Wilkerson's Theorem.
(3) For the second half, show that a λ-ring with $\psi^n = Id$ for all n is necessarily $\mathbf{Z}$-torsion-free.
(4) With the $\mathbf{Z}$-torsion-free condition, the only λ-operations that can give rise to the Adams operations $\psi^n = Id$ are the ones in (5.2). This implies that the ring is a binomial ring.

Lemma 5.5. *In a binomial ring R, one has*

$$a^p \equiv a \pmod{pR} \tag{5.3}$$

for $a \in R$ and prime p.

Proof. Since $\mathbf{Z}/p$ is the splitting field of the polynomial $x^p - x$, we have

$$x^p - x \equiv \prod_{i=0}^{p-1}(x-i) \pmod{p\mathbf{Z}}. \tag{5.4}$$

(See, for example, Chapter V, Section 5 in [Lang (2002)].) It follows that for $a \in R$ we have

$$\begin{aligned} a^p - a &\equiv a(a-1)\cdots(a-(p-1)) \pmod{pR} \\ &= p!\binom{a}{p} \in p!R \\ &\equiv 0 \pmod{pR}. \end{aligned}$$

This proves the Lemma. □

Lemma 5.6. *Let R be a λ-ring in which all the Adams operations are equal to the identity map. Then*

$$a(a-1)\cdots(a-(n-1)) = n!\lambda^n(a) \tag{5.5}$$

for $a \in R$ and $n \geq 1$.

Proof. Since $\psi^n = Id$ for all n, we have

$$\begin{aligned} \frac{-at}{1+t} &= \sum_{n\geq 1}(-1)^n at^n = \psi_{-t}(a) = -t\frac{d}{dt}(\log \lambda_t(a)) \\ &= \frac{-\sum_{n\geq 1} n\lambda^n(a)t^n}{\sum_{n\geq 0}\lambda^n(a)t^n}. \end{aligned}$$

Clearing out the denominators, we have

$$(1+t)\sum_{n\geq 1} n\lambda^n(a)t^n = at\sum_{n\geq 0}\lambda^n(a)t^n,$$

The coefficients of t^n give

$$n\lambda^n(a) + (n-1)\lambda^{n-1}(a) = a\lambda^{n-1}(a),$$

which is equivalent to

$$n\lambda^n(a) = (a-(n-1))\lambda^{n-1}(a). \tag{5.6}$$

Now (5.5) is trivially true when $n = 1$. The higher cases are proved by induction using (5.6):

$$\begin{aligned} n!\lambda^n(a) &= (n-1)!\cdot n\lambda^n(a) \\ &= (n-1)!\cdot(a-(n-1))\lambda^{n-1}(a) \quad \text{by } (5.6) \\ &= (a-(n-1))\cdot a(a-1)\cdots(a-(n-2)) \end{aligned}$$

by the induction hypothesis. This finishes the proof of the Lemma. □

The first half of Theorem 5.3 is contained in the following result.

Proposition 5.7. *Let R be a binomial ring. Then R has the unique structure of a λ-ring in which all the Adams operations are equal to the identity map. Moreover, the λ-operations are given by*

$$\lambda^n(a) = \binom{a}{n} = \frac{a(a-1)\cdots(a-(n-1))}{n!} \tag{5.7}$$

for $a \in R$ and $n \geq 1$.

Proof. Equip R with the trivial ψ-ring (Definition 3.44) structure $\psi^n = Id\colon R \to R$, in which each ψ^n is the identity map on R. Thus, R is a $\mathbf{Z}$-torsion-free ψ-ring satisfying (5.3). By Wilkerson's Theorem 3.54 there exists a unique λ-ring structure on R in which each Adams operation ψ^n is equal to the identity map on R.

To prove (5.7), pick an element $a \in R$ and an integer $n \geq 1$. By Lemma 5.6 we know that the λ-operations in R satisfy

$$a(a-1)\cdots(a-(n-1)) = n!\lambda^n(a).$$

Since R is $\mathbf{Z}$-torsion-free by definition, there is a well-defined element

$$\lambda^n(a) = \frac{a(a-1)\cdots(a-(n-1))}{n!} \in R.$$

This proves (5.7). □

Before we proceed with the proof of Theorem 5.3, let us record the following result, which will provide another description of binomial rings.

Corollary 5.8. *Let R be a $\mathbf{Z}$-torsion-free ring satisfying $a^p \equiv a \pmod{pR}$ for $a \in R$ and prime p. Then R has a unique λ-ring structure in which all the Adams operations are equal to the identity map.*

Proof. Use the first paragraph of the proof of Proposition 5.7. □

To prove the second half of Theorem 5.3, we need the following preliminary results.

Lemma 5.9. *Let R be a ring, and let $f = 1 + a_1t + a_2t^2 + \cdots$ be a strict power series over R (Definition 2.1). Suppose that*

$$f^p = \underbrace{f \times \cdots \times f}_{p \text{ factors}} = 1$$

for some prime p. Then we have $pa_k = 0$ for $1 \leq k < p$ and $a_1^p + pa_p = 0$.

Proof. Since we are only concerned about $a_1, \ldots, a_p$, we will work modulo t^{p+1}. By the Multinomial Theorem (Section 1.2 in [Stanley (1997)]) we have

$$\begin{aligned}
1 &\equiv (1 + a_1t + \cdots + a_pt^p)^p \pmod{t^{p+1}} \\
&\equiv \sum_{k_0+\cdots+k_p=p} \frac{p!}{k_0!\cdots k_p!}(a_1t)^{k_1}\cdots(a_pt^p)^{k_p} \pmod{t^{p+1}} \\
&\equiv 1 + pa_1t + \left(\binom{p}{2}a_1^2 + pa_2\right)t^2 + \cdots + (a_1^p + \cdots + pa_p)t^p \pmod{t^{p+1}}.
\end{aligned}$$

In the above sum, suppose that j is the largest positive integer for which $k_j \neq 0$ for a typical summand. Then the corresponding coefficient in front is

$$\frac{p!}{k_0!\cdots k_j!} = \frac{(p-k_j)!}{k_0!\cdots k_{j-1}!}\cdot\binom{p}{k_j}, \tag{5.8}$$

which is divisible by p, unless $k_j = p$. This last case only happens when $j = 1$, and the corresponding term in f^p is

$$a_1^p t^p. \tag{5.9}$$

Moreover, the coefficient of t^i in f^p has the form

$$pa_i + (\text{terms involving } a_1, \ldots, a_{i-1}). \tag{5.10}$$

In particular, we can start with the observation $pa_1 = 0$ from the above expansion. For $1 < k < p$, we use induction, (5.8), and (5.10) to conclude that $pa_k = 0$. This proves the first assertion of the Lemma.

For the second assertion, we use the first assertion, (5.8), (5.9), and (5.10) to conclude that the coefficient of t^p in f^p is $a_1^p + pa_p$. Since the t^p term in f^p is zero by assumption, we have proved the second assertion. □

Proposition 5.10. *Let R be a λ-ring in which all the Adams operations are equal to the identity map. Then R is $\mathbf{Z}$-torsion-free.*

Proof. Using the prime factorizations of integers, it clearly suffices to show that, if

$$pa = 0 \tag{5.11}$$

for some prime p and $a \in R$, then $a = 0$. To prove this, we compute as follows:

$$\begin{aligned} a^p - a &= a(a-1)\cdots(a-(p-1)) \quad \text{by (5.4) and (5.11)} \\ &= p!\lambda^p(a) \quad \text{by Lemma 5.6.} \end{aligned} \tag{5.12}$$

Now by the assumption (5.11), we have

$$1 = \lambda_t(pa) = \underbrace{\lambda_t(a) \times \cdots \times \lambda_t(a)}_{p \text{ times}},$$

since λ_t is additive (Proposition 1.13 (3)) and the addition in $\Lambda(R)$ is multiplication of strict power series (2.4). Therefore, we can apply Lemma 5.9 with $f = \lambda_t(a) = 1 + \sum_{n\geq 1} \lambda^n(a)t^n$. We obtain

$$a^p + p\lambda^p(a) = 0,$$

or equivalently,

$$p\lambda^p(a) = -a^p. \tag{5.13}$$

Putting (5.13) back into (5.12), we have

$$\begin{aligned} a^p - a &= (p-1)! \cdot p\lambda^p(a) \\ &= -(p-1)!a^p \\ &= a^p. \end{aligned} \tag{5.14}$$

The last equality follows from

$$(p-1)! \equiv -1 \pmod{p}$$

and (5.11). Therefore, we conclude from (5.14) that $a = 0$, as desired. □

We now finish the proof of Theorem 5.3.

Proof. [Proof of Theorem 5.3] The first assertion is contained in Proposition 5.7.

For the converse, let R be a λ-ring in which all the Adams operations are equal to the identity map. By Proposition 5.10 R is **Z**-torsion-free. Thus, it remains to show (5.2). Pick an element $a \in R$ and an integer $n \geq 1$. By Lemma 5.6 we know that

$$a(a-1)\cdots(a-(n-1)) = n!\lambda^n(a).$$

Since R is **Z**-torsion-free, we have a well-defined element

$$\lambda^n(a) = \frac{a(a-1)\cdots(a-(n-1))}{n!} \in R.$$

This shows (5.2) and that R is a binomial ring. □

At this point we have two addition descriptions of a binomial ring. We record them as follows.

Corollary 5.11. *Let R be a ring. Then the following statements are equivalent.*

(1) R is a binomial ring.
(2) R is $\mathbf{Z}$-torsion-free and satisfies $a^p \equiv a \pmod{pR}$ for $a \in R$ and prime p.
(3) R is a λ-ring in which all the Adams operations are equal to the identity map.

Proof. We have (1) ⇒ (2) by Lemma 5.5, (2) ⇒ (3) by Corollary 5.8, and (3) ⇒ (1) by Theorem 5.3. □

In §5.4 we will add another characterization to the above list (Theorem 5.34).

Here is another immediate consequence of Theorem 5.3, which is due to [Wilkerson (1982)].

Corollary 5.12. *The rings $\mathbf{Z}_{(p)}$ of p-local integers and $\hat{\mathbf{Z}}_p$ of p-adic integers are binomial rings. In particular, they are $\mathbf{Z}$-torsion-free λ-rings with*

$$\lambda^n(a) = \frac{a(a-1)\cdots(a-(n-1))}{n!}$$

for $n \geq 1$ and $a \in \mathbf{Z}_{(p)}$ or $\hat{\mathbf{Z}}_p$.

Proof. We know from Corollary 3.59 and Corollary 3.63 that $\mathbf{Z}_{(p)}$ and $\hat{\mathbf{Z}}_p$ are λ-rings in which all the Adams operations are equal to the identity map. Therefore, by Theorem 5.3 $\mathbf{Z}_{(p)}$ and $\hat{\mathbf{Z}}_p$ are binomial rings, which are necessarily **Z**-torsion-free. The assertion about λ^n follows from (5.2). □

Binomial rings have another nice property. To describe it we need the following definition.

Definition 5.13. A ring R is said to be *radical* if, for any $n \geq 1$,

$$a^n \in pR \quad \text{if and only if} \quad a \in pR \tag{5.15}$$

for p prime and $a \in R$.

Of course, the "if" part of this definition is always true. For example, the ring $\mathbf{Z}$ of integers is radical.

The following consequence of Theorem 5.3 provides more examples of rings that are radical and is also due to [Wilkerson (1982)].

Corollary 5.14. *Every binomial ring is radical.*

Proof. Let R be a binomial ring. By Theorem 5.3 R is a λ-ring in which all the Adams operations are equal to the identity map. Pick an element $a \in R$ with $a^n \in pR$ for some prime p and integer $n \geq 1$. We want to show that a lies in pR. We have

$$\begin{aligned} a &= \psi^{p^n}(a) \\ &\equiv a^{p^n} \pmod{pR} \quad \text{by (3.4)} \\ &\equiv a^n a^{p^n - n} \pmod{pR} \\ &\equiv 0 \pmod{pR}. \end{aligned}$$

This proves that R is radical. □

Applying this Corollary to the known examples of binomial rings (Example 5.2 and Corollary 5.12), we obtain the following result.

Corollary 5.15. *The rings $\mathbf{Z}_{(p)}$ of p-local integers, $\hat{\mathbf{Z}}_p$ of p-adic integers, and all the $\mathbf{Q}$-algebras are radical.*

5.2 Integer-Valued Polynomials

In this section we begin our discussion of rings of integer-valued polynomials. We show that a ring of integer-valued polynomials is a binomial ring (Theorem 5.21). A general reference for integer-valued polynomials is [Cahen and Chabert (1997)].

In the following sections, we will establish a $\mathbf{Z}$-module basis for such a ring, a result that is due to [Wilkerson (1982)]. Moreover, we will use rings of integer-valued polynomials to give another characterization of binomial rings in §5.4.

Definition 5.16. Let X be a set of variables, and let $\mathbf{Q}[X]$ denote the ring of polynomials in X with rational coefficients. Define the set of *integer-valued polynomials* as

$$\mathrm{Int}(\mathbf{Z}^X) = \left\{ f \in \mathbf{Q}[X] \colon f\left(\mathbf{Z}^X\right) \subseteq \mathbf{Z} \right\}.$$

It is a subring of $\mathbf{Q}[X]$ and is called the *ring of integer-valued polynomials in X*.

In this definition, the condition $f\left(\mathbf{Z}^X\right) \subseteq \mathbf{Z}$ is interpreted as follows. First, $\mathbf{Z}^X$ is the set of functions $\underline{n} \colon X \to \mathbf{Z}$, which is a choice of an integer for each $x \in X$. We can evaluate f at any such $\underline{n}$ by replacing each variable $x \in X$ with the integer $\underline{n}(x)$ in f. The defining condition for $f \in \mathbf{Q}[X]$ to be an integer-valued polynomial is that $f(\underline{n}) \in \mathbf{Z}$ for every $\underline{n} \in \mathbf{Z}^X$. In other words, f takes on integer values when evaluated at integers.

Example 5.17. If X is the empty set, then we have $\text{Int}(\mathbf{Z}^{\varnothing}) = \mathbf{Z}$, the ring of integers.

Example 5.18. In the polynomial ring $\mathbf{Q}[x]$ in one variable, the binomial symbol

$$\binom{x}{n} = \frac{x(x-1)\cdots(x-(n-1))}{n!} \in \mathbf{Q}[x]$$

for any $n \geq 0$ is an integer-valued polynomial. For example, we have

$$\binom{x}{0} = 1, \binom{x}{1} = x, \binom{x}{2} = \frac{x(x-1)}{2},$$
$$\binom{x}{3} = \frac{x(x-1)(x-2)}{3!}.$$

Note that for any integer m,

$$\binom{m}{n} = \lambda^n(m) = \text{coefficient of } t^n \text{ in } (1+t)^m,$$

where λ^n is the λ-operation in $\mathbf{Z}$ (Example 1.15 and Corollary 2.11). So, for example, we have

$$\binom{-1}{2} = 1 \quad \text{and} \quad \binom{-2}{2} = 3.$$

Example 5.19. For any prime p and integer n, we know that $n^p \equiv n \pmod{p}$. Therefore, in the polynomial ring $\mathbf{Q}[x]$ in one variable,

$$F_p(x) = \frac{x^p - x}{p}$$

is an integer-valued polynomial, called a *Fermat polynomial.* Any iterated composition

$$F_p \circ \cdots \circ F_p(x) = F_p(\cdots F_p(F_p(x)))$$

of $F_p(x)$ is also an integer-valued polynomial. For example, when $p = 3$, we have

$$F_3(x) = \frac{x^3 - x}{3}.$$

Moreover, we can write $F_3(x)$ as a $\mathbf{Z}$-linear combination of the binomial symbols (Example 5.18) as follows:

$$\begin{aligned} F_3(x) &= \frac{x(x-1)((x-2)+3)}{3} \\ &= \frac{x(x-1)(x-2)}{3} + x(x-1) \\ &= 2\binom{x}{3} + 2\binom{x}{2}. \end{aligned}$$

As we will see below (Theorem 5.28), this is not a coincident. In fact, the binomial symbols $\binom{x}{n}$ $(n \geq 0)$ form a $\mathbf{Z}$-module basis for the ring $\text{Int}(\mathbf{Z}^{\{x\}})$ of integer-valued polynomials in one variable x.

Example 5.20. In the polynomial ring $\mathbf{Q}[x, y]$ in two variables, the product

$$\binom{x}{n}\binom{y}{m}$$

is an integer-valued polynomial for any $n, m \geq 0$. It follows that any $\mathbf{Z}$-linear combination

$$\sum_{i=1}^{r} k_i \binom{x}{n_i}\binom{y}{m_i}$$

of such products is also an integer-valued polynomial.

Here is the main result of this section.

Theorem 5.21. *Let* $\mathrm{Int}(\mathbf{Z}^X)$ *be the ring of integer-valued polynomials for some set* X *of variables. Then* $\mathrm{Int}(\mathbf{Z}^X)$ *is a binomial ring (Definition 5.1).*

Proof. Since $\mathrm{Int}(\mathbf{Z}^{\varnothing}) = \mathbf{Z}$, which is a binomial ring, we may assume that X is not empty. Since $\mathrm{Int}(\mathbf{Z}^X)$ is a subring of $\mathbf{Q}[X]$ by definition, it is clearly $\mathbf{Z}$-torsion-free. To see that $\mathrm{Int}(\mathbf{Z}^X)$ is closed under taking binomial symbols, let f be an element in $\mathrm{Int}(\mathbf{Z}^X)$ and $n \geq 1$. We want to show that

$$\binom{f}{n} = \frac{f(f-1)\cdots(f-(n-1))}{n!} \in \mathbf{Q}[X]$$

is an integer-valued polynomial. Pick an element $\underline{m} \in \mathbf{Z}^X$. Then $f(\underline{m}) \in \mathbf{Z}$, which implies that

$$\begin{aligned}\binom{f}{n}(\underline{m}) &= \frac{f(\underline{m})(f(\underline{m})-1)\cdots(f(\underline{m})-(n-1))}{n!} \\ &= \binom{f(\underline{m})}{n} \in \mathbf{Z},\end{aligned}$$

as desired. □

As we saw in the last section, binomial rings have many good properties. In particular, all of those properties of binomial rings must hold for rings of integer-valued polynomials.

Corollary 5.22. *Let* $\mathrm{Int}(\mathbf{Z}^X)$ *be the ring of integer-valued polynomials for some set* X *of variables. Then the following statements hold.*

(1) The ring $\mathrm{Int}(\mathbf{Z}^X)$ *is a* λ*-ring in which all the Adams operations are equal to the identity map, and*

$$\lambda^n(f) = \binom{f}{n} = \frac{f(f-1)\cdots(f-(n-1))}{n!}$$

for $n \geq 1$ *and* $f \in \mathrm{Int}(\mathbf{Z}^X)$.

(2) The congruence

$$f^p \equiv f \pmod{p\,\mathrm{Int}(\mathbf{Z}^X)}$$

holds for $f \in \mathrm{Int}(\mathbf{Z}^X)$ *and* p *prime.*

(3) The ring $\mathrm{Int}(\mathbf{Z}^X)$ *is radical (Definition 5.13).*

Proof. Apply Theorem 5.3, Lemma 5.5, and Corollary 5.14 to the binomial ring $\mathrm{Int}(\mathbf{Z}^X)$. □

5.3 A Basis for Integer-Valued Polynomials

The purpose of this section is to establish a $\mathbf{Z}$-module basis for the ring $\mathrm{Int}(\mathbf{Z}^X)$ of integer-valued polynomials over a set X (Theorem 5.28). This result (at least when X is finite) is due to [Wilkerson (1982)]. In the next section, we will use this result to give another characterization of a general binomial ring in terms of rings of integer-valued polynomials.

Let X be a non-empty set of variables and $\mathrm{Int}(\mathbf{Z}^X)$ be the corresponding ring of integer-valued polynomials.

Definition 5.23. By a *multi-index* we mean an element

$$I = (i_x)_{x \in X} \in \bigoplus_{x \in X} \mathbf{Z}_{\geq 0},$$

in which $\mathbf{Z}_{\geq 0}$ denotes the set of non-negative integers. Given such a multi-index I, define the *generalized binomial symbol*

$$\binom{X}{I} = \prod_{x \in X} \binom{x}{i_x} \in \mathrm{Int}(\mathbf{Z}^X). \tag{5.16}$$

This makes sense because each binomial symbol $\binom{x}{i_x}$ is an integer-valued polynomial, and only finitely many of the $i_x \in \mathbf{Z}_{\geq 0}$ are non-zero. If $J = (j_x)$ is another multi-index, then we define

$$\binom{J}{I} = \prod_{x \in X} \binom{j_x}{i_x} \in \mathbf{Z}.$$

Example 5.24. For the trivial multi-index $I = (0)$, i.e., each $i_x = 0$, we have

$$\binom{X}{(0)} = \prod_{x \in X} \binom{x}{0} = 1.$$

In general, if $i_{x_1}, \ldots, i_{x_n}$ are the only non-zero entries in I, then

$$\binom{X}{I} = \prod_{k=1}^{n} \binom{x_k}{i_{x_k}},$$

a finite product of binomial symbols.

Example 5.25. Note that if n and m are integers such that $n > m \geq 0$, then $\binom{m}{n} = 0$. Therefore, if $I = (i_x)$ and $J = (j_x)$ are two multi-indexes, then we have

$$\binom{J}{I} = \prod_{x \in X} \binom{j_x}{i_x} = \begin{cases} 1 & \text{if } I = J, \\ 0 & \text{if } j_x < i_x \text{ for some } x \in X. \end{cases}$$

The previous example suggests that there is a natural order for the multi-indexes, at least when X is finite. This is the lexicographical order.

Definition 5.26. Let $X = \{x_1, \ldots, x_n\}$ be a *finite* non-empty set of variables. We consider a multi-index corresponding to X as an n-tuple

$$I = (i_1, \ldots, i_n) \in \bigoplus_{k=1}^{n} \mathbf{Z}_{\geq 0}.$$

Define the *lexicographical order* on the set of multi-indexes as follows. Let $J = (j_1, \ldots, j_n)$ be another multi-index. Then we set $I > J$ if and only if $i_1 = j_1, \ldots, i_k = j_k$, and $i_{k+1} > j_{k+1}$ for some k with $0 \leq k \leq n-1$.

Example 5.27. When $X = \{x_1, \ldots, x_n\}$ consists of n variables, the smallest multi-index under the lexicographical order is $0 = (0, \ldots, 0)$, which has 0 in every entry. If $I = (i_k)$ and $J = (j_k)$ are two multi-indexes, then

$$I > J \quad \text{implies} \quad \binom{J}{I} = 0.$$

The converse is *not* true. For example, we have $\binom{2}{1}\binom{0}{1} = 0$, but $(2, 0) > (1, 1)$ under the lexicographical order.

Here is the main result of this section, which says that the generalized binomial symbols form a $\mathbf{Z}$-module basis for the ring of integer-valued polynomials.

Theorem 5.28. *Let X be a non-empty set of variables. Then the generalized binomial symbols* (5.16)

$$\left\{\binom{X}{I} : I = (i_x) \in \bigoplus_{x \in X} \mathbf{Z}_{\geq 0} \text{ is a multi-index}\right\}$$

form a $\mathbf{Z}$-module basis for the ring $\mathrm{Int}(\mathbf{Z}^X)$ *of integer-valued polynomials.*

The proof of this Theorem consists of the following steps:

(1) Show that the generalized binomial symbols form a $\mathbf{Q}$-module basis for the polynomial ring $\mathbf{Q}[X]$.
(2) For an integer-valued polynomial f, write it as a $\mathbf{Q}$-linear combination of the generalized binomial symbols. Show that the coefficients must all be integers.

We begin with the assertion about the $\mathbf{Q}$-module basis.

Lemma 5.29. *Let $X = \{x_1, \ldots, x_n\}$ be a finite non-empty set of variables. Then the generalized binomial symbols*

$$\left\{\binom{X}{I} : I = (i_1, \ldots, i_n) \in \bigoplus_{k=1}^{n} \mathbf{Z}_{\geq 0}\right\}$$

are linearly independent over $\mathbf{Q}$.

Proof. Suppose to the contrary that the generalized binomial symbols are linearly dependent over $\mathbf{Q}$. Thus, there exists a linear relation

$$q_1 \binom{X}{I_1} + \cdots + q_r \binom{X}{I_r} = 0 \tag{5.17}$$

for some $r \geq 1$, in which each $q_k \in \mathbf{Q}$ is non-zero and $I_1, \ldots, I_r$ are distinct. Order the multi-indexes in (5.17) lexicographically (Definition 5.26). By reordering the indexes if necessary, we may assume without loss of generality that $I_1 < I_2 < \cdots < I_r$. In particular, we have

$$\binom{I_1}{I_k} = \begin{cases} 1 & \text{if } k = 1, \\ 0 & \text{if } 2 \leq k \leq r. \end{cases}$$

Now evaluate (5.17) at $X = I_1$:

$$0 = q_1 \binom{I_1}{I_1} + q_2 \binom{I_1}{I_2} + \cdots + q_r \binom{I_1}{I_r} = q_1.$$

This contradicts the condition that q_1 is non-zero. Therefore, the linear relation (5.17) cannot exist, and the generalized binomial symbols are linearly independent over $\mathbf{Q}$. □

Lemma 5.30. *Let $X = \{x_1, \ldots, x_n\}$ be a finite non-empty set of variables. Then the generalized binomial symbols*

$$\left\{ \binom{X}{I} : I = (i_1, \ldots, i_n) \in \bigoplus_{k=1}^{n} \mathbf{Z}_{\geq 0} \right\}$$

span $\mathbf{Q}[X]$ over $\mathbf{Q}$.

Proof. Since $\binom{X}{(0)} = 1$, it suffices to consider polynomials of positive degrees. We do induction on the degrees of the polynomials. Suppose by induction that the generalized binomial symbols span the polynomials in $\mathbf{Q}[X]$ of degree $< m$. Pick a polynomial $f \in \mathbf{Q}[X]$ of degree m. We want to show that f can be written as a $\mathbf{Q}$-linear combination of the generalized binomial symbols.

If $I = (i_1, \ldots, i_n)$ is a multi-index, we use the shorthand $x^I = x_1^{i_1} \cdots x_n^{i_n}$. There are only finitely many non-zero monomials in f of degree m (1.3). Thus, f has the form

$$f = \left(\sum_{I \in S} q_I x^I \right) + (\text{terms of degree} < m) \tag{5.18}$$

for some non-empty finite set S, in which each monomial x^I has degree m and each $q_I \in \mathbf{Q}$ is non-zero. Observe that for each variable $x \in X$, we have

$$\begin{aligned} k! \binom{x}{k} &= x(x-1) \cdots (x - (k-1)) \\ &= x^k + (\text{terms in } x \text{ of degree} < k). \end{aligned} \tag{5.19}$$

For each multi-index $I = (i_1, \ldots, i_n) \in S$, we use the shorthand $I! = (i_1!) \cdots (i_n!) \in \mathbf{Z}$. Using (5.19) we have

$$\begin{aligned} I!\binom{X}{I} &= \prod_{j=1}^{n} i_j!\binom{x_j}{i_j} \\ &= \prod_{j=1}^{n} \left(x_j^{i_j} + (\text{terms in } x_j \text{ of degree} < i_j)\right) \\ &= x^I + (\text{terms of degree} < m). \end{aligned}$$

Putting this back into (5.18), it follows that the polynomial

$$f_1 = f - \sum_{I \in S} q_I \cdot I!\binom{X}{I} \in \mathbf{Q}[X] \tag{5.20}$$

has degree $< m$. By the induction hypothesis, we can write f_1 as a $\mathbf{Q}$-linear combination of the generalized binomial symbols. This implies that f can be written as a $\mathbf{Q}$-linear combination of the generalized binomial symbols, which finishes the induction. The proof of the Lemma is complete. □

Note that the proof of Lemma 5.30 provides an inductive procedure for writing a polynomial $f \in \mathbf{Q}[X]$ as a $\mathbf{Q}$-linear combination of the generalized binomial symbols. One starts with the highest degree terms in f and considers the difference f_1 (5.20). Now apply the same procedure to f_1, and so forth. In each step, the degree of the resulting polynomial is at least one less than that of the previous step. As we will see below in the proof of Theorem 5.28, if $f \in \mathrm{Int}(\mathbf{Z}^X)$ then the resulting coefficients must all be integers.

Combining Lemmas 5.29 and 5.30 we obtain the following result.

Proposition 5.31. *Let X be any non-empty set of variables. Then the generalized binomial symbols*

$$\left\{\binom{X}{I} : I = (i_x) \in \bigoplus_{x \in X} \mathbf{Z}_{\geq 0} \text{ is a multi-index}\right\}$$

form a $\mathbf{Q}$-module basis for $\mathbf{Q}[X]$.

Proof. Since each multi-index is non-zero in only finitely many entries, any $\mathbf{Q}$-linear relation

$$\sum q_I \binom{X}{I} = 0 \tag{5.21}$$

among the generalized binomial symbols involves only finitely many variables in X. Therefore, any such relation (5.21) is actually happening in some subring $\mathbf{Q}[Y]$ for some finite subset $Y \subseteq X$. By Lemma 5.29 this relation must be trivial, i.e., each q_I is zero. Therefore, the generalized binomial symbols are linearly independent over $\mathbf{Q}$.

To see that the generalized binomial symbols span $\mathbf{Q}[X]$, suppose that f is a polynomial in $\mathbf{Q}[X]$. There are only finitely many variables in X that actually occur in f. Thus, f lies in some subring $\mathbf{Q}[Y]$ for some finite subset $Y \subseteq X$. By Lemma 5.30 f can be written as a $\mathbf{Q}$-linear combination of the generalized binomial symbols. □

We can now finish the proof of Theorem 5.28.

Proof. [Proof of Theorem 5.28] Since the generalized binomial symbols are linearly independent over $\mathbf{Q}$ (Proposition 5.31), they are also linearly independent over $\mathbf{Z}$. Moreover, any $\mathbf{Z}$-linear combination of the generalized binomial symbols lies in $\mathrm{Int}(\mathbf{Z}^X)$. It remains to show that the generalized binomial symbols span $\mathrm{Int}(\mathbf{Z}^X)$ over $\mathbf{Z}$.

Pick an integer-valued polynomial $f \in \mathrm{Int}(\mathbf{Z}^X)$ with $f \neq 0$. Since f involves only finitely many variables in X, we may assume without loss of generality that X is finite, say, $X = \{x_1, \ldots, x_n\}$. By Lemma 5.30 we can write

$$f = \sum_{j=1}^{r} q_j \binom{X}{I_j} \tag{5.22}$$

for some $r \geq 1$ with each $q_j \in \mathbf{Q}$ and $I_1, \ldots, I_r$ distinct. Since $f \in \mathrm{Int}(\mathbf{Z}^X)$ we have

$$f(I) = \sum_{j=1}^{r} q_j \binom{I}{I_j} \in \mathbf{Z}$$

for any multi-index I.

To finish the proof, it suffices to show that each coefficient q_j lies in $\mathbf{Z}$. Order the multi-indexes in the sum (5.22) lexicographically (Definition 5.26). By reordering the indexes if necessary, we may assume without loss of generality that $I_1 < I_2 < \cdots < I_r$. In particular, we have

$$\binom{I_k}{I_l} = \begin{cases} 1 & \text{if } k = l, \\ 0 & \text{if } k < l. \end{cases} \tag{5.23}$$

Using (5.23) we evaluate f at $X = I_1$, which gives $f(I_1) = q_1 \in \mathbf{Z}$. Inductively, suppose that $q_1, \ldots, q_{k-1} \in \mathbf{Z}$. We want to show that q_k is an integer. We evaluate f at $X = I_k$ using (5.23), which gives

$$\begin{aligned} f(I_k) &= \sum_{j=1}^{r} q_j \binom{I_k}{I_j} \in \mathbf{Z} \\ &= q_1 \binom{I_k}{I_1} + \cdots + q_{k-1} \binom{I_k}{I_{k-1}} + q_k. \end{aligned}$$

Since the sum of the first $k-1$ terms is an integer by the induction hypothesis, we conclude that q_k is also an integer. Therefore, by induction every q_j $(1 \leq j \leq r)$ is an integer, as desired.

This finishes the proof of Theorem 5.28. □

Example 5.32. A $\mathbf{Z}$-module basis for the ring $\text{Int}(\mathbf{Z}^{\{x\}})$ of integer-valued polynomials in one variable is given by the set of binomial symbols

$$\left\{\binom{x}{n} = \frac{x(x-1)\cdots(x-(n-1))}{n!} : n \geq 0\right\}.$$

In the two-variable case, a $\mathbf{Z}$-module basis for $\text{Int}(\mathbf{Z}^{\{x,y\}})$ is given by the set of products

$$\left\{\binom{x}{n}\binom{y}{m} : n, m \geq 0\right\}.$$

In the r-variable case, a $\mathbf{Z}$-module basis for $\text{Int}(\mathbf{Z}^{\{x_1,\ldots,x_r\}})$ is given by the set of products

$$\left\{\binom{x_1}{n_1}\cdots\binom{x_r}{n_r} : n_1, \ldots, n_r \geq 0\right\}.$$

5.4 Binomial Rings as Integer-Valued Polynomials

Recall from Theorem 5.3 that a λ-ring in which all the Adams operations are equal to the identity map is equivalent to a binomial ring. The main purpose of this section is to characterize binomial rings as $\mathbf{Z}$-torsion-free homomorphic images of rings of integer-valued polynomials (Theorem 5.34). This result is due to [Elliott (2006)].

Given two objects A and B in a category $\mathcal{C}$, recall that the symbol $\mathcal{C}(A, B)$ denotes the set of morphisms from A to B. The categories of rings and sets are denoted by **Ring** and **Set**, respectively.

Let X be a non-empty set of variables, and let R be a binomial ring. The following result shows that any function $g\colon X \to R$ of sets can be extended uniquely to a ring homomorphism $\text{Int}(\mathbf{Z}^X) \to R$, whose restriction to X is equal to g. In other words, the ring $\text{Int}(\mathbf{Z}^X)$ is the *free binomial ring* on the set X.

Theorem 5.33. *Let X be a non-empty set of variables, and let R be a binomial ring. Consider the map*

$$\rho\colon \mathbf{Ring}(\text{Int}(\mathbf{Z}^X), R) \to \mathbf{Set}(X, R)$$

defined by $\rho(\varphi)(x) = \varphi(x)$ for $\varphi \in \mathbf{Ring}(\text{Int}(\mathbf{Z}^X), R)$ and $x \in X$. Then ρ is a bijection.

Proof. First we show that ρ is injective. Observe that for a ring homomorphism $\varphi \in \mathbf{Ring}(\text{Int}(\mathbf{Z}^X), R)$, $x \in X$, and $n \geq 1$, we have

$$\begin{aligned} n!\varphi\binom{x}{n} &= \varphi\left(x(x-1)\cdots(x-(n-1))\right) \\ &= \varphi(x)(\varphi(x)-1)\cdots(\varphi(x)-(n-1)) \qquad (5.24) \\ &= n!\binom{\varphi(x)}{n}. \end{aligned}$$

Since R is $\mathbf{Z}$-torsion-free, it follows that $\varphi\binom{x}{n} = \binom{\varphi(x)}{n}$. By Theorem 5.28 the ring homomorphism φ is uniquely determined by

$$\begin{aligned}\varphi\binom{X}{I} &= \prod_{x\in X}\varphi\binom{x}{i_x}\\ &= \prod_{x\in X}\binom{\varphi(x)}{i_x},\end{aligned}$$

where I runs through the multi-indexes. Thus, φ is completely determined by its actions on the polynomials $x \in X$. This implies that ρ is injective.

Next we show that ρ is surjective. Let $g: X \to R$ be a function. We want to show that $g = \rho(\varphi)$ for some ring homomorphism $\varphi \in \mathbf{Ring}(\mathrm{Int}(\mathbf{Z}^X), R)$. The function g extends uniquely to a ring homomorphism $\phi: \mathbf{Q}[X] \to R \otimes \mathbf{Q}$ with $\phi(x) = g(x) \in R$ for each $x \in X$. Using Theorem 5.28 and the definition of a binomial ring, it follows that the restriction

$$\varphi = \phi|_{\mathrm{Int}(\mathbf{Z}^X)}: \mathrm{Int}(\mathbf{Z}^X) \to R \otimes \mathbf{Q}$$

has image in R. Moreover, we have $\rho(\varphi)(x) = \phi(x) = g(x)$ for any $x \in X$. This proves that ρ is surjective. □

We can now state the main result of this section, which gives a characterization of binomial rings in terms of integer-valued polynomials.

Theorem 5.34. *A ring R is a binomial ring if and only if the following two conditions hold:*

(1) R is $\mathbf{Z}$-torsion-free.
(2) R is the homomorphic image of a ring $\mathrm{Int}(\mathbf{Z}^X)$ of integer-valued polynomials for some set X.

Proof. For the "if" part, suppose that the two stated conditions are satisfied. So there exists a surjective ring homomorphism $\varphi: \mathrm{Int}(\mathbf{Z}^X) \to R$ for some set X. We want to show that R is closed under taking binomial symbols. Pick an element $r \in R$ and an integer $n \geq 1$. Since φ is surjective, we have $\varphi(f) = r$ for some $f \in \mathrm{Int}(\mathbf{Z}^X)$. Moreover, $\mathrm{Int}(\mathbf{Z}^X)$ is a binomial ring (Theorem 5.21), so we have $\binom{f}{n} \in \mathrm{Int}(\mathbf{Z}^X)$. Then it follows exactly as in (5.24) (with f in place of x) that

$$\varphi\binom{f}{n} = \binom{\varphi(f)}{n} = \binom{r}{n} \in R.$$

Thus, we have shown that R is a binomial ring.

For the "only if" part, suppose that R is a binomial ring. Then R is $\mathbf{Z}$-torsion-free by definition. It remains to prove the second condition. Consider the set $X = \{x_r : r \in R\}$, which consists of a variable x_r for each element $r \in R$, and the function $g: X \to R$ defined by $g(x_r) = r$ for $r \in R$. By Theorem 5.33 there exists a unique ring homomorphism $\varphi: \mathrm{Int}(\mathbf{Z}^X) \to R$ such that $\varphi(x_r) = g(x_r) = r$ for $r \in R$. In particular, φ is surjective, proving the second condition. □

Combining Corollary 5.11 and Theorem 5.34 we obtain the following four equivalent descriptions of the same class of rings.

Corollary 5.35. *Let R be a ring. Then the following four statements are equivalent.*

(1) R is a binomial ring.
(2) R is $\mathbf{Z}$-torsion-free and satisfies $a^p \equiv a \pmod{pR}$ for $a \in R$ and prime p.
(3) R is a λ-ring in which all the Adams operations are equal to the identity map.
(4) R is $\mathbf{Z}$-torsion-free and is the homomorphic image of a ring $\mathrm{Int}(\mathbf{Z}^X)$ of integer-valued polynomials for some set X.

5.5 Knutson's Theorem

The purpose of this section is to prove Knutson's Theorem [Knutson (1975)]. It states that a binomial ring with a certain kind of generating subset must be isomorphic to $\mathbf{Z}$. This result has immediate applications to topological K-theory and representation theory. In fact, it implies that the only finite group G whose representation ring $R(G)$ is a binomial ring is the trivial group. Similarly, if the topological K-theory $K(X)$ is a binomial ring, then $K(X)$ must be isomorphic to $\mathbf{Z}$. (However, it is not necessarily the case that X is contractible.)

Recall from Theorem 5.3 that a binomial ring is equivalently a λ-ring in which all the Adams operations are equal to the identity map. Here is the main result of this section.

Theorem 5.36 (Knutson's Theorem). *Let R be a binomial ring. Assume that there exists a subset $A \subseteq R$ satisfying the following conditions:*

(1) The subset A generates R as an abelian group.
(2) Every element $a \in A$ has finite dimension (Definition 1.11).
(3) The subset A is closed under the λ-operations λ^n $(n \geq 1)$.
(4) If $a \in A$ has dimension one, then a is an invertible element in R.

Then R is isomorphic to the ring $\mathbf{Z}$ of integers.

The plan to prove Knutson's Theorem is to start with an element $a \in A$ of dimension, say, n. Then we show that $a = n$, so A consists of only non-negative integers. This implies the result because A generates R as an abelian group.

We need the following preliminary result.

Lemma 5.37. *In a binomial ring R, one has $n\lambda^n(x) = x\lambda^{n-1}(x-1)$ for $x \in R$ and $n \geq 1$.*

Proof. From Theorem 5.3 we know that R is a $\mathbf{Z}$-torsion-free λ-ring in which

$$\lambda^n(x) = \binom{x}{n} = \frac{x(x-1)\cdots(x-(n-1))}{n!}.$$

Therefore, we have

$$\begin{aligned} n\lambda^n(x) &= \frac{x(x-1)\cdots(x-(n-1))}{(n-1)!} \\ &= x\cdot\frac{(x-1)((x-1)-1)\cdots((x-1)-(n-2))}{(n-1)!} \\ &= x\lambda^{n-1}(x-1), \end{aligned}$$

as desired. □

Proof. [Proof of Theorem 5.36] Pick an element $a \in A$. By hypothesis a has finite dimension, say, n. In particular, we have

$$\lambda^{n+1}(a) = 0. \tag{5.25}$$

Then $b = \lambda^n(a) \in A$ has dimension one by Corollary 1.48. Since b has dimension one, we have $\psi^n(b) = b^n$ for all $n \geq 1$ by Proposition 3.5. However, since R is a binomial ring, as a λ-ring its Adams operations are all equal to the identity map (Theorem 5.3). Thus, we have $b = \psi^2(b) = b^2$, i.e., b is idempotent. By hypothesis the element b is invertible in R, so $bc = 1$ for some element $c \in R$. It follows that

$$1 = bc = b^2c = b. \tag{5.26}$$

Now observe that

$$\begin{aligned} \lambda_t(a) &= \lambda_t(1+(a-1)) \\ &= \lambda_t(1)\lambda_t(a-1) \quad \text{by (1.8)} \\ &= (1+t)\lambda_t(a-1). \end{aligned}$$

Collecting the coefficients of t^n, we obtain

$$\lambda^n(a) = \lambda^{n-1}(a-1) + \lambda^n(a-1). \tag{5.27}$$

Using the information above, we compute as follows:

$$\begin{aligned} a &= ab \quad \text{by (5.26)} \\ &= a\lambda^n(a) \\ &= a\lambda^{n-1}(a-1) + a\lambda^n(a-1) \quad \text{by (5.27)} \\ &= n\lambda^n(a) + (n+1)\lambda^{n+1}(a) \quad \text{by Lemma 5.37} \\ &= nb + (n+1)\cdot 0 \quad \text{by (5.25)} \\ &= n \quad \text{by (5.26)}. \end{aligned}$$

Since $a \in A$ is arbitrary, we conclude that A consists of non-negative integers. Therefore, R is isomorphic to $\mathbf{Z}$ because A generates R as an abelian group. □

Now we apply Knutson's Theorem 5.36 to the representation ring of a finite group, which is always a λ-ring (Example 1.17).

Corollary 5.38. *Let G be a finite group. If the representation ring $R(G)$ of G is a binomial ring, then G is the trivial group.*

Proof. Take $A \subseteq R(G)$ to be the subset consisting of the actual representations of G (as opposed to virtual representations). Then clearly A satisfies the hypotheses of Knutson's Theorem 5.36. Thus, we have $R(G) \cong \mathbf{Z}$, so the only actual representations of G are the trivial representations. If G has a non-identity element, then the regular representation of G is non-trivial. Therefore, we conclude that G is the trivial group. □

Next we apply Knutson's Theorem 5.36 to topological K-theory (Example 1.16).

Corollary 5.39. *Let X be a connected compact Hausdorff space. Suppose that the K-theory $K(X)$ of X is a binomial ring. Then we have an isomorphism $K(X) \cong \mathbf{Z}$ of rings.*

Proof. Take $A \subseteq K(X)$ to be the subset consisting of the actual vector bundles on X (as opposed to virtual bundles). From the basic theory of vector bundles, the set A satisfies the hypotheses of Knutson's Theorem 5.36. The result follows. □

Note that in Corollary 5.39 we did not claim that X is contractible. Indeed, there are non-contractible spaces whose K-theory is trivial, i.e., isomorphic to $\mathbf{Z}$.

5.6 Necklace Rings

The purpose of this section is to show that, for a binomial ring R, the universal λ-ring $\Lambda(R)$ of R (Theorem 2.6) has yet another interesting description. Namely, it is isomorphic to the necklace ring $Nr(R)$ of R (Theorem 5.42), which was first introduced by [Metropolis and Rota (1983)]. This result is due to [Elliott (2006)], although it also appeared in a different form in [Metropolis and Rota (1983)].

5.6.1 *Primitive necklaces*

Before we give the definition of the necklace ring, let us describe the results that led Metropolis and Rota to define it.

Suppose that you have a large bag of beads and that the beads come in x different colors. You want to make a necklace with the beads in the bag by arranging n beads in a circle. We can denote a necklace by a word

$$w = a_1 a_2 \cdots a_n,$$

where $a_1, \ldots, a_n$ are n beads from the bag. Since this is a necklace, if the beads are cyclically permutated, we consider the result to be the same necklace. For example, we have

$$\begin{aligned} w &= a_1 a_2 \cdots a_n \\ &= a_n a_1 \cdots a_{n-1} \\ &= a_{n-1} a_n a_1 \cdots a_{n-2}, \end{aligned}$$

and so forth.

A necklace w is called *non-primitive* if there is a string v of d beads, with d dividing n and $d < n$, such that w is constructed by repeating the string v (n/d) times, i.e.,

$$w = \underbrace{v \cdots v}_{n/d \text{ times}} .$$

Otherwise, it is called *primitive.* The natural question arises: How many different primitive necklaces can be constructed? The answer is a well known fact in combinatorics [Cameron (1995); Stanley (2001)]. Namely, there are

$$M(x, n) = \frac{1}{n} \sum_{d|n} \mu\left(\frac{n}{d}\right) x^d \tag{5.28}$$

primitive necklaces of length n that can be constructed using a set of beads with x colors. Here μ is the Möbius function, which is defined as follows. If a positive integer m has prime factorization $m = p_1^{e_1} \cdots p_n^{e_n}$, then

$$\mu(m) = \begin{cases} 1 & \text{if } m = 1, \\ 0 & \text{if } e_i > 1 \text{ for some } i, \\ (-1)^n & \text{if } e_1 = \cdots = e_n = 1. \end{cases}$$

We will not need to use the expression (5.28) for $M(x, n)$.

5.6.2 *Necklace polynomials*

In (5.28) we can consider x as a variable. In this case, $M(x, n)$ becomes a polynomial of degree n in $\mathbf{Q}[x]$ and is called the *necklace polynomial.* Using purely combinatorial counting arguments, Metropolis and Rota proved in [Metropolis and Rota (1983)] that

$$M(xy, n) = \sum_{[i,j]=n} (i, j) M(x, i) M(y, j), \tag{5.29}$$

where

$$[i, j] = \text{the least common multiple of } i \text{ and } j,$$
$$(i, j) = \text{the greatest common divisor of } i \text{ and } j.$$

This property of the necklace polynomial is the main motivation of the definition of the necklace ring.

Definition 5.40. Let R be a ring. The *necklace ring* $Nr(R)$ of R is the ring with underlying set $Nr(R) = \prod_{n=1}^{\infty} R$ and entry-wise addition. As usual an element in $Nr(R)$ is denoted by a sequence $\mathbf{a} = (a_1, a_2, \ldots)$. The multiplication in $Nr(R)$ is defined by

$$(\mathbf{a} \times \mathbf{b})_n = \sum_{[i,j]=n} (i, j) a_i b_j, \tag{5.30}$$

where $(\mathbf{a} \times \mathbf{b})_n$ denotes the nth entry in the product $\mathbf{a} \times \mathbf{b}$.

Clearly $Nr(R)$ is an abelian group under its addition with additive zero $(0, 0, \ldots)$. Its multiplicative identity is $(1, 0, 0, \ldots)$. The other ring axioms (associativity of $\times$ and the distributive property) are not hard to check and are left as exercises for the reader at the end of this chapter.

5.6.3 *Relations with* $\Lambda(R)$

The necklace ring should remind the reader of the ring $\mathbf{W}(R)$ of big Witt vectors on R (§4.1). In fact, the underlying sets of $Nr(R)$ and $\mathbf{W}(R)$ are equal by definition. The role of the Witt polynomial w_n (4.1) is now played by the necklace polynomial $M(x, n)$ (5.28).

Since the ring $\mathbf{W}(R)$ is canonically isomorphic to the universal λ-ring $\Lambda(R)$ via the Artin-Hasse Exponential isomorphism (Theorem 4.16), one might expect that the necklace ring $Nr(R)$ is closely related to $\Lambda(R)$ as well. Indeed, these two rings are isomorphic, at least for the class of binomial rings.

In order to define this isomorphism, we first need the following preliminary result.

Lemma 5.41. *Let R be a binomial ring. Then every strict power series (Definition 2.1) $f = 1 + r_1 t + r_2 t^2 + \cdots$ over R can be written uniquely as $f = \prod_{n=1}^{\infty} (1 - (-1)^n t^n)^{a_n}$ for some elements $a_1, a_2, \ldots \in R$.*

Notice the similarity between this Lemma and Lemma 4.14. Here we interpret the factor $(1 - (-1)^n t^n)^{a_n}$ using (5.1) as the strict power series

$$\begin{aligned}(1 - (-1)^n t^n)^{a_n} &= 1 + \sum_{i \geq 1} \binom{a_n}{i} (-1)^{(n+1)i} (t^n)^i \\ &= 1 + \sum_{i \geq 1} (-1)^{(n+1)i} \frac{a_n(a_n - 1) \cdots (a_n - (i-1))}{i!} t^{ni} \\ &= 1 + (-1)^{n+1} a_n t^n + \binom{a_n}{2} t^{2n} + \cdots .\end{aligned}$$

Of course, this only makes sense when R is a binomial ring.

Proof. [Proof of Lemma 5.41] Expanding the product we obtain:

$$\begin{aligned}&\prod_{n=1}^{\infty} (1 - (-1)^n t^n)^{a_n} \\ &\quad = (1 + t)^{a_1} (1 - t^2)^{a_2} (1 + t^3)^{a_3} \cdots \\ &\quad = 1 + a_1 t + \left(-a_2 + \binom{a_1}{2} \right) t^2 + \cdots + \\ &\qquad \left((-1)^{n+1} a_n + \mathbf{Z}\text{-polynomial in } \binom{a_i}{j}, i < n \right) t^n + \cdots .\end{aligned}$$

Using this expansion we can solve for a_n inductively in terms of $r_1, \ldots, r_n$: $a_1 = r_1$, $a_2 = -r_2 + \binom{r_1}{2}$, and so forth. □

With Lemma 5.41 we can now state the main result of this section.

Theorem 5.42. *Let R be a binomial ring. Then the map $E_{Nr}\colon \Lambda(R) \to Nr(R)$ defined by*

$$E_{Nr}\left(\prod_{n=1}^{\infty}(1-(-1)^n t^n)^{a_n}\right) = (a_1, a_2, \ldots)$$

is a ring isomorphism.

Observe the similarity between the map E_{Nr} and the Artin-Hasse Exponential isomorphism (Theorem 4.16). As noted above, this result is due to Elliott [Elliott (2006)]. However, our isomorphism E_{Nr} is defined slightly differently from that in [Elliott (2006)], which goes in the opposite direction $E\colon Nr(R) \to \Lambda(R)$. Moreover, Elliott's E is defined with the signs $E(\mathbf{a}) = \prod_{n=1}^{\infty}(1-t^n)^{-a_n}$, which requires using $(1-t)^{-1} = 1 + t + t^2 + \cdots \in \Lambda(R)$ (as opposed to $(1+t)$) as the multiplicative identity. Furthermore, our proof takes full advantage of the Cartesian product ring of R, which (we hope) makes the proof more transparent.

The plan to prove Theorem 5.42 is to construct an injective ring homomorphism $\phi\colon Nr(R) \to R^{\mathbf{N}}$ from $Nr(R)$ to the Cartesian product ring $R^{\mathbf{N}}$. Then we observe that the composition $\phi E_{Nr}\colon \Lambda(R) \to R^{\mathbf{N}}$ is also an injective ring homomorphism. These two facts together imply that E_{Nr} is a ring homomorphism. The bijectivity of E_{Nr} is easy. We need some preliminary results.

Recall from (4.20) the ring homomorphisms $z_n\colon \Lambda(R) \to R$ defined by the generating function

$$z_{-t}(f) = -t\frac{d}{dt}(\log f) = \frac{-tf'}{f},$$

where $z_t(f) = \sum_{n\geq 1} z_n(f)t^n$. The map z_n corresponds to the Witt polynomial w_n via the Artin-Hasse Exponential isomorphism (Theorem 4.21).

Lemma 5.43. *Let R be a binomial ring. Then we have*

$$z_k\left((1-(-1)^n t^n)^a\right) = \begin{cases} 0 & \text{if } n \text{ does not divide } k, \\ na & \text{if } n \text{ divides } k \end{cases} \tag{5.31}$$

for all $k, n \geq 1$ and $a \in R$.

Proof. From the definition of z_{-t}, we have

$$\begin{aligned} z_{-t}\left((1-(-1)^n t^n)^a\right) &= -t\frac{d}{dt}\left(\log(1-(-1)^n t^n)^a\right) \\ &= \frac{(-1)^n nat^n}{1-(-1)^n t^n} \\ &= (-1)^n nat^n\left(1 + (-1)^n t^n + t^{2n} + (-1)^{3n} t^{3n} + \cdots\right) \\ &= (-1)^n nat^n + nat^{2n} + (-1)^{3n} nat^{3n} + \cdots. \end{aligned}$$

Therefore, $z_k\left((1-(-1)^n t^n)^a\right)$ is given as in (5.31). $\square$

Recall the Cartesian product ring $R^{\mathbf{N}} = \prod_{n=1}^{\infty} R$ with entry-wise addition and multiplication (3.20). For a binomial ring R, consider the map $\phi\colon Nr(R) \to R^{\mathbf{N}}$ defined by

$$\phi\left((a_1, a_2, \ldots)\right)_k = \sum_{n|k} na_n, \tag{5.32}$$

where $\phi\left((a_1, a_2, \ldots)\right)_k \in R$ denotes the kth entry in $\phi\left((a_1, a_2, \ldots)\right) \in R^{\mathbf{N}}$. In particular, for an element $(\ldots, 0, a_n, 0, \ldots) \in Nr(R)$ that is non-zero only in the nth entry, ϕ is given by

$$\phi\left((\ldots, 0, a_n, 0, \ldots)\right)_k = \begin{cases} 0 & \text{if } n \text{ does not divide } k, \\ na_n & \text{if } n \text{ divides } k. \end{cases} \tag{5.33}$$

In the case $n = 1$, this says that ϕ preserves the multiplicative identity, since $\phi((1, 0, 0, \ldots)) = (1, 1, 1, \ldots)$ by (5.33).

Lemma 5.44. *Let R be a binomial ring. Then the map $\phi\colon Nr(R) \to R^{\mathbf{N}}$ defined in (5.32) is a ring homomorphism.*

Proof. Since the additions in both $Nr(R)$ and $R^{\mathbf{N}}$ are defined entry-wise, it is clear that ϕ is additive and preserves the additive zero. We already observed that ϕ preserves the multiplicative identity. Thus, it remains to show that ϕ is multiplicative.

Observe that every element in $Nr(R)$ is uniquely an infinite sum

$$(a_1, a_2, \ldots) = \sum_{n \geq 1} (\ldots, 0, a_n, 0, \ldots)$$

of elements with at most one non-zero entry, one for each $n \geq 1$. Using the additivity of ϕ, to prove the multiplicativity of ϕ it suffices to consider elements with only one non-zero entry. Let $\mathbf{a} = (\ldots, 0, a_n, 0, \ldots)$ and $\mathbf{b} = (\ldots, 0, b_m, 0, \ldots)$ be two such elements in $Nr(R)$, where a_n and b_m are in the nth entry and the mth entry, respectively. Then by the definition of the necklace multiplication (5.30) we have

$$(\mathbf{a} \times \mathbf{b})_{[n,m]} = (n, m) a_n b_m,$$

and all the other entries in $\mathbf{a} \times \mathbf{b}$ are equal to zero. Therefore, we have

$$\phi\left(\mathbf{a} \times \mathbf{b}\right)_k = \begin{cases} 0 & \text{if } [n, m] \nmid k, \\ [n, m] \cdot (n, m) a_n b_m = nm a_n b_m & \text{if } [n, m] \mid k \end{cases}$$

by (5.33).

On other other hand, both $\phi(\mathbf{a})$ and $\phi(\mathbf{b})$ are defined as in (5.33). Since the multiplication in the Cartesian product ring $R^{\mathbf{N}}$ is defined entry-wise, we know that

$$(\phi(\mathbf{a}) \times \phi(\mathbf{b}))_k = \phi(\mathbf{a})_k \times \phi(\mathbf{b})_k = 0$$

if either n does not divide k or m does not divide k, i.e., if $[n,m]$ does not divide k. Likewise, we have

$$\begin{aligned}(\phi(\mathbf{a}) \times \phi(\mathbf{b}))_k &= \phi(\mathbf{a})_k \times \phi(\mathbf{b})_k \\ &= (na_n) \cdot (mb_m) \\ &= nma_nb_m\end{aligned}$$

if both n and m divide k, i.e., if $[n,m]$ divides k. Therefore, we conclude that ϕ is multiplicative and is a ring homomorphism. □

We are now ready to prove Theorem 5.42.

Proof. [Proof of Theorem 5.42] By Lemma 5.41 the map E_{Nr} is well-defined and is a bijection. The map E_{Nr} preserves the additive zero and the multiplicative identity, since $E_{Nr}(1) = (0,0,0,\ldots)$ and $E_{Nr}(1+t) = (1,0,0,\ldots)$. Recall that the addition $+_\Lambda$ in the universal λ-ring $\Lambda(R)$ is multiplication of strict power series (2.4). Since the addition in $Nr(R)$ is defined entry-wise, it follows that E_{Nr} is additive. It remains to show that E_{Nr} is multiplicative.

Now consider the map $z\colon \Lambda(R) \to R^{\mathbf{N}}$ defined by

$$z(f) = (z_1(f), z_2(f), z_3(f), \ldots)$$

for $f \in \Lambda(R)$. Since each z_n is a ring homomorphism from $\Lambda(R)$ to R (Theorem 4.21), the map z is also a ring homomorphism. Moreover, from the expression (4.22) of $z_n(f)$ and the assumption that R is $\mathbf{Z}$-torsion-free, it follows that z is injective.

We claim that the diagram

$$\begin{array}{ccc}\Lambda(R) & \xrightarrow{E_{Nr}} & Nr(R) \\ & {\scriptstyle z}\searrow \quad \swarrow{\scriptstyle \phi} & \\ & R^{\mathbf{N}} & \end{array} \tag{5.34}$$

is commutative. Indeed, for $n, k \geq 1$ and $a_n \in R$ we have

$$\begin{aligned}\phi\left(E_{Nr}\left((1-(-1)^n t^n)^{a_n}\right)\right)_k &= \phi\left((\ldots,0,a_n,0,\ldots)\right)_k \\ &= \begin{cases} 0 & \text{if } n \text{ does not divide } k, \\ na_n & \text{if } n \text{ divides } k \end{cases} \quad \text{by (5.33)} \\ &= z_k\left((1-(-1)^n t^n)^{a_n}\right).\end{aligned}$$

The last equality follows from Lemma 5.43. Thus, the maps ϕE_{Nr} and z agree on elements of the form $(1-(-1)^n t^n)^{a_n} \in \Lambda(R)$. Now ϕ, E_{Nr}, and z are all additive, so it follows from Lemma 5.41 and the definition of $+_\Lambda$ (2.4) that $\phi E_{Nr} = z$ on $\Lambda(R)$.

The commutativity of (5.34) and the facts that E_{Nr} is a bijection and that z is injective imply that ϕ is injective. Using the multiplicativity of z and ϕ (Lemma 5.44) and the injectivity of ϕ, we conclude that E_{Nr} is multiplicative as well.

This finishes the proof of Theorem 5.42. □

Corollary 5.45. *Let R be a binomial ring. Then the necklace ring $Nr(R)$ has a canonical λ-ring structure.*

Proof. The ring $\Lambda(R)$ is a λ-ring (Theorem 2.6). Thus, $Nr(R)$ becomes a λ-ring by structure transport via the isomorphism E_{Nr}. In other words, we define

$$\lambda^n = E_{Nr}\lambda^n E_{Nr}^{-1} \colon Nr(R) \to Nr(R),$$

where the λ^n on the right is the nth λ-operation on $\Lambda(R)$. □

5.7 Exercises

(1) Write the Fermat polynomial $F_p(x) = (x^p - x)/p$ in terms of the $\mathbf{Z}$-module basis $\{\binom{x}{n}\}_{n\geq 0}$ (Example 5.32) for $p = 5, 7$, and 11.

(2) Let X be a non-empty set of variables. Prove that the map

$$\varphi\colon \bigotimes_{x\in X} \mathrm{Int}(\mathbf{Z}^{\{x\}}) \to \mathrm{Int}(\mathbf{Z}^X)$$

defined by $\varphi(x) = x$ for $x \in X$ is a ring isomorphism.

(3) Let R be a binomial ring without zero divisiors, and let S be a multiplicative subset in R. Show that the localization $S^{-1}R$ is also a binomial ring.

(4) Let $\{R_i\}$ be a non-empty set of rings with i running through some non-empty set S. Prove: The product ring $\prod_i R_i$ is binomial if and only if each R_i is binomial.

(5) Let $f \in \mathbf{Z}[x_1, \ldots, x_k]$ be a polynomial in k variables. Suppose that for some positive integer m, one has $f(n_1, \ldots, n_k) \equiv 0 \pmod{m\mathbf{Z}}$ for all $n_1, \ldots, n_k \in \mathbf{Z}$. Let R be a binomial ring, and let $r_1, \ldots, r_k \in R$. Show that $f(r_1, \ldots, r_k) \equiv 0 \pmod{mR}$. In other words, any integer congruence that is satisfied by all the integers must also be satisfied by elements in a binomial ring.

(6) Let R be a binomial ring. Show that $Nr(R)$ is actually a ring.

(7) Show that the necklace ring construction gives rise to a functor from the category of rings to itself.

(8) Show that the isomorphism $E_{Nr}\colon \Lambda(R) \to Nr(R)$ of Theorem 5.42 gives rise to a natural isomorphism of functors $E_{Nr}\colon \Lambda \to Nr$ from the category of binomial rings to the category of rings.

Chapter 6

Filtered λ-Rings, I

This is the first of two chapters in which we study filtered λ-rings. In §6.1 we introduce the notion of a *filtered λ-ring*, which is a λ-ring together with a decreasing sequence of λ-ideals. For example, the γ-filtration of an augmented λ-ring R makes it into a filtered λ-ring. In particular, the K-theory $K(X)$ of a pointed space X and the representation ring $R(G)$ of a group G are both filtered λ-rings.

In §6.2 we consider the following quotient power series or polynomial rings

$$R = \frac{\mathbf{Z}[[x_1, \ldots, x_n]]}{\langle x_1^{e_1}, \ldots, x_n^{e_n} \rangle} \quad \text{or} \quad \frac{\mathbf{Z}[x_1, \ldots, x_n]}{\langle x_1^{e_1}, \ldots, x_n^{e_n} \rangle}.$$

Here $x_1, \ldots, x_n$ are variables, $1 \leq e_i \leq \infty$ for each i, and at least one $e_j < \infty$. We set $x_k^{\infty} = 0$ if $e_k = \infty$. The symbol $\langle x_1^{e_1}, \ldots, x_n^{e_n} \rangle$ denotes the ideal generated by $x_1^{e_1}, \ldots, x_n^{e_n}$. The variables x_i are given a certain fixed filtration, say, d. It is shown that such a filtered ring R admits *uncountably many isomorphism classes* of filtered λ-ring structures. For example, this statement applies to the filtered rings

$$\frac{\mathbf{Z}[x]}{\langle x^2 \rangle}, \quad \frac{\mathbf{Z}[x_1, x_2]}{\langle x_1^4 \rangle}, \quad \frac{\mathbf{Z}[[x_1, x_2, x_3]]}{\langle x_1^5, x_3^{100} \rangle}, \quad \text{and} \quad \frac{\mathbf{Z}[[x_1, \ldots, x_n]]}{\langle x_n^3 \rangle}.$$

Not surprisingly, this result is an application of Wilkerson's Theorem. Actually the uncountability assertion also holds for the power series rings $\mathbf{Z}[[x_1, \ldots, x_n]]$ for $n \geq 1$. However, the only proof of this statement [Yau (2004b)] that the author knows of uses a substantial amount of homotopy theory. We will not discuss the proof of this case in details in this book.

For a general filtered ring R, we denote by $\lambda(R)$ the (possibly empty) set of filtered λ-ring structures on R. In §6.3 we describe explicitly all the isomorphism classes in the uncountable set $\lambda(\mathbf{Z}[x]/(x^2))$ (Corollary 6.17). These isomorphism classes are not hard to describe. In fact, every such isomorphism class corresponds to a sequence (a_p) of integers, one for each prime p, in which $a_p \in p\mathbf{Z}$.

In §6.4 we describe explicitly all the isomorphism classes in the uncountable set $\lambda(\mathbf{Z}[x]/(x^3))$. Compared to the case of $\lambda(\mathbf{Z}[x]/(x^2))$, the isomorphism classes in $\lambda(\mathbf{Z}[x]/(x^3))$ are more complicated to describe.

The references for this Chapter are [Yau (2004a, 2006)]. The reader is referred to [Yau (2003a,b, 2004b, 2005)] for applications of filtered λ-rings to homotopy theory.

6.1 Filtered λ-Rings

In this section we introduce filtered λ-rings and give some examples. We will also discuss a filtered version of Wilkerson's Theorem, which will be used in later sections many times.

6.1.1 *Filtered rings*

We begin with the notion of a filtered ring.

Definition 6.1. By a *filtered ring* we mean a ring R together with a decreasing sequence $R = I_0 \supseteq I_1 \supseteq I_2 \supseteq \cdots$ of ideals. The ideals I_n are called the *filtration ideals.* Such a filtered ring is denoted by either R or $(R, \{I_n\})$. A *filtered ring homomorphism* (or *isomorphism*) $\varphi\colon R \to S$ between two filtered rings is a ring homomorphism (or isomorphism) of the underlying rings that preserves the filtration ideals, in the sense that $\varphi\left(I_n^R\right) \subseteq I_n^S$ for all n. Here we denote by I_n^R the nth filtration ideal in R.

Note that there is a category $\mathbf{Ring}^f$ whose objects are filtered rings and whose morphisms are filtered ring homomorphisms.

Example 6.2. Let p be a prime. The ring of integers $\mathbf{Z}$ is a filtered ring when equipped with the filtration ideals $I_n = p^n\mathbf{Z}$ for $n \geq 0$. This is called the *p-adic filtration.* Note that these filtration ideals are the ones used in the construction of the p-adic integers (Examples 3.60 and 3.61).

Example 6.3. Consider the power series ring $R = A[[x]]$ in one variable over some subring $A \subseteq \mathbf{Q}$ (e.g., $A = \mathbf{Z}$ or $\mathbf{Z}_{(p)}$ for some prime p). Let I_n be the principal ideal $I_n = \langle x^n \rangle = x^n R$ generated by x^n. Then R is a filtered ring with these I_n as its filtration ideals. This is call the *x-adic filtration.* A filtered ring homomorphism $\varphi\colon A[[x]] \to S$ from R to any filtered ring S is uniquely determined by the element $\varphi(x) \in I_1^S$. All of these hold if R is replaced by either the polynomial ring $A[x]$ or the truncated polynomial ring $A[x]/\langle x^e \rangle$ for some positive integer $e \geq 2$.

Example 6.4. More generally, consider the power series ring $R = A[[x_1, \ldots, x_n]]$ in n variables for some $n \geq 1$, where A is some subring of the rationals $\mathbf{Q}$. Assign to each variable x_i a fixed positive filtration, say, $d > 0$. Non-zero elements in A have filtration 0, and $0 \in A$ has filtration ∞. A monomial $ax_1^{k_1} \cdots x_n^{k_n} \in R$ with $a \neq 0$ in A has filtration $d(k_1 + \cdots + k_n)$. Let I_k be the ideal in R generated by elements of filtration at least k. Thus, elements in I_k have the form

$$f = \sum a_{k_1,\ldots,k_n} x_1^{k_1} \cdots x_n^{k_n}$$

with $a_{k_1,\ldots,k_n} \in A$ and $d(k_1 + \cdots + k_n) \geq k$. Then R is a filtered ring with I_k ($k \geq 0$) as its filtration ideals. This is called the *adic filtration* of R. A filtered ring homomorphism $\varphi\colon A[[x_1, \ldots, x_n]] \to S$ to any filtered ring S is uniquely determined by the n elements $\varphi(x_1), \ldots, \varphi(x_n) \in I_d^S$. All of these still hold if R is replaced by the quotients

$$R = \frac{A[[x_1, \ldots, x_n]]}{\langle x_1^{e_1}, \ldots, x_n^{e_n} \rangle} \quad \text{or} \quad \frac{A[x_1, \ldots, x_n]}{\langle x_1^{e_1}, \ldots, x_n^{e_n} \rangle}$$

for some exponents e_i with $2 \leq e_i \leq \infty$, where we set $x_i^\infty = 0$ if $e_i = \infty$.

6.1.2 *Filtered λ-rings*

Next we define a filtered version of λ-rings.

Definition 6.5. By a *filtered λ-ring* we mean a λ-ring R that is also a filtered ring in which each filtration ideal I_n is closed under λ^k for $k \geq 1$ (i.e., I_n is a λ-ideal). A *filtered λ-homomorphism* (or *λ-isomorphism*) $\varphi \colon R \to S$ between two filtered λ-rings is a filtered ring homomorphism (or isomorphism) of the underlying filtered rings that is also a λ-homomorphism (Definition 1.25).

In other words, a filtered λ-ring is a λ-ring together with a decreasing filtration of λ-ideals. A filtered λ-homomorphism is a ring homomorphism that preserves the filtration ideals and is compatible with the λ-operations.

Example 6.6. Let R be an augmented λ-ring. Recall from Definition 1.30 that this means that there is a λ-homomorphism $\varepsilon \colon R \to \mathbf{Z}$. Then there is a γ-filtration $\{I_n\}$ (Definition 3.28). The nth filtration ideal I_n is generated as an abelian group by the products $\gamma^{n_1}(a_1) \cdots \gamma^{n_r}(a_r)$, where each $a_i \in I = \ker \varepsilon$ (the augmentation ideal) and $n_1 + \cdots + n_r \geq n$. The γ-filtration makes R into a filtered λ-ring, since each γ-filtration ideal I_n is a λ-ideal (Proposition 3.27 (4)). If $\varphi \colon R \to S$ is a λ-homomorphism between augmented λ-rings that preserves the augmentation ideal, then φ is a filtered λ-homomorphism.

Example 6.7. The representation ring $R(G)$ of a group G is an augmented λ-ring (Example 1.32). Therefore, as in Example 6.6, $R(G)$ becomes a filtered λ-ring when equipped with its γ-filtration. If $\varphi \colon G \to H$ is a group homomorphism, then the induced map $\varphi^* \colon R(H) \to R(G)$ on the representation rings is a λ-homomorphism that preserves the augmentation ideal. In particular, φ^* is a filtered λ-homomorphism.

Example 6.8. The K-theory of a pointed space X is an augmented λ-ring (Example 1.31). Therefore, with its γ-filtration, $K(X)$ is a filtered λ-ring. If $\varphi \colon X \to Y$ is a map of pointed spaces, then the induced map $\varphi^* \colon K(Y) \to K(X)$ on K-theory is a λ-homomorphism that preserves the augmentation ideal. Thus, φ^* is a filtered λ-homomorphism.

Example 6.9. The K-theory of a pointed space X has another natural filtered λ-ring structure. Suppose that X has the homotopy type of a CW complex. Fix such a CW structure on X. For $n \geq 1$ let X_n be the nth skeleton in this CW structure. Then the inclusion map $i_n\colon X_n \hookrightarrow X$ induces a λ-homomorphism $i_n^*\colon K(X) \to K(X_n)$ on K-theory. For $n \geq 1$ set $J_n = \ker i_n^* \subseteq K(X)$ and $J_0 = K(X)$. Then $K(X)$ becomes a filtered λ-ring when equipped with the filtration ideals J_n $(n \geq 0)$. We leave it as an exercise for the reader to check that this filtered λ-ring structure on $K(X)$ is, up to isomorphism, independent of the choice of the CW structure on X.

Proposition 6.10. *Let R be a filtered λ-ring. Then every Adams operation ψ^n on R is a filtered λ-homomorphism.*

Proof. We already know that ψ^n is a λ-homomorphism (Theorem 3.6). It remains to show that ψ^n preserves the filtration ideals. Pick an element $a \in I_k$ for some $k \geq 1$. Then we have $\lambda^j(a) \in I_k$ for all $j \geq 1$, since I_k is a λ-ideal. Therefore, in the Newton Formula (Theorem 3.10)

$$\psi^n(a) - \lambda^1(a)\psi^{n-1}(a) + \cdots + (-1)^{n-1}\lambda^{n-1}(a)\psi^1(a) = (-1)^{n+1}n\lambda^n(a),$$

all the terms except $\psi^n(a)$ lie in I_k. Thus, $\psi^n(a)$ must also lie in I_k, and ψ^n preserves the filtration ideals. □

6.1.3 *Filtered Wilkerson's Theorem*

For the purpose of constructing filtered λ-ring structures, it is convenient to have a filtered version of Wilkerson's Theorem 3.54. Recall that the Adams operations ψ^n for $n \geq 1$ are uniquely determined by the subset consisting of the ψ^p for p primes (3.3).

Theorem 6.11 (Filtered Wilkerson's Theorem). *Let R be a $\mathbf{Z}$-torsion-free filtered ring with filtration ideals I_n $(n \geq 0)$. Suppose that for each prime p there is a filtered ring endomorphism $\psi^p\colon R \to R$ satisfying the following conditions:*

(1) $\psi^p\psi^q = \psi^q\psi^p$ for any primes p and q.
(2) $\psi^p(a) \equiv a^p \pmod{pR}$ for $a \in R$ and p prime.
(3) For $a \in R$ and $n, m > 0$, $ma \in I_n$ implies $a \in I_n$.

Then there exists a unique filtered λ-ring structure on R with the given ψ^p as its Adams operations.

Proof. Define the maps $\psi^1 = Id\colon R \to R$ and

$$\psi^n = (\psi^{p_1})^{e_1} \cdots (\psi^{p_k})^{e_k}\colon R \to R \tag{6.1}$$

if $n > 1$ has prime factorization $n = p_1^{e_1} \cdots p_k^{e_k}$. By the commutativity of the ψ^p for p primes, all the ψ^n are well-defined. Moreover, it is immediate that $\psi^n\psi^m = \psi^{nm}$ and that each ψ^n is a filtered ring homomorphism on R. Thus, R is a $\mathbf{Z}$-torsion ψ-ring (Definition 3.44) that satisfies the hypotheses of Wilkerson's Theorem 3.54.

Therefore, there exists a unique λ-ring structure on R with the ψ^n (6.1) as its Adams operations.

It remains to show that this is, in fact, a *filtered* λ-ring structure on R. In other words, we need to show that

$$a \in I_n \quad \text{implies} \quad \lambda^k(a) \in I_n \tag{6.2}$$

for $k \geq 1$. Now if $a \in I_n$, then $\psi^j(a) \in I_n$ for all $j \geq 1$ because ψ^j is a filtered ring homomorphism. Therefore, in the Newton Formula (Theorem 3.10)

$$\psi^k(a) - \lambda^1(a)\psi^{k-1}(a) + \cdots + (-1)^{k-1}\lambda^{k-1}(a)\psi^1(a) = (-1)^{k+1}k\lambda^k(a),$$

the left-hand side lies in I_n. Thus, we have $k\lambda^k(a) \in I_n$. By hypothesis this implies that $\lambda^k(a) \in I_n$. This finishes the proof of the Theorem. □

The above filtered version of Wilkerson's Theorem takes on a particularly simple form when applied to power series and polynomial rings.

Corollary 6.12. *Consider the filtered ring*

$$R = \frac{A[[x_1, \ldots, x_n]]}{\langle x_1^{e_1}, \ldots, x_n^{e_n} \rangle} \quad \text{or} \quad \frac{A[x_1, \ldots, x_n]}{\langle x_1^{e_1}, \ldots, x_n^{e_n} \rangle}$$

with the adic filtration (Example 6.4), for any subring $A \subseteq \mathbf{Q}$, $n \geq 1$, and exponents e_i with $2 \leq e_i \leq \infty$. Then a filtered λ-ring structure on R with Adams operations ψ^k $(k \geq 1)$ is uniquely determined by the power series (or polynomials) $\psi^p(x_i) \in R$ for $1 \leq i \leq n$ and p prime satisfying the following three conditions:

(1) The constant term in each $\psi^p(x_i)$ is 0.
(2) $\psi^p(\psi^q(x_i)) = \psi^q(\psi^p(x_i))$ for any primes p and q.
(3) $\psi^p(x_i) \equiv x_i^p \pmod{pR}$.

Proof. If R has a filtered λ-ring structure, then its Adams operations are filtered λ-homomorphisms (Proposition 6.10). Since each x_i has positive filtration, so must $\psi^p(x_i)$. Thus, $\psi^p(x_i)$ has 0 constant term. The other two conditions are properties of Adams operations (Theorem 3.7).

Conversely, suppose that the three stated conditions are satisfied. We want to show that there exists a unique filtered λ-ring structure on R whose Adams operations ψ^p agree with the given power series (or polynomials) when applied to x_i. For each prime p, since $\psi^p(x_i)$ $(1 \leq i \leq n)$ has 0 constant term, it extends uniquely to a filtered ring homomorphism $\psi^p \colon R \to R$ satisfying

$$\psi^p\left(\sum a_{k_1,\ldots,k_n} x_1^{k_1} \cdots x_n^{k_n}\right) = \sum a_{k_1,\ldots,k_n} \psi^p(x_1)^{k_1} \cdots \psi^p(x_n)^{k_n}$$

Since R is clearly $\mathbf{Z}$-torsion-free, it suffices to show that these filtered ring homomorphisms ψ^p satisfy the hypotheses of the Filtered Wilkerson's Theorem 6.11.

We know that $\psi^p\psi^q = \psi^q\psi^p$ for any primes p and q, since they agree on the ring generators $x_1, \ldots, x_n$ by hypothesis. Moreover, the pth power map on R is linear modulo pR. Therefore, the hypothesis $\psi^p(x_i) \equiv x_i^p \pmod{pR}$ implies that

$\psi^p(r) \equiv r^p \pmod{pR}$ for all $r \in R$. Finally, let I_k denote the kth filtration ideal in R. For $r \in R$ and $n, m > 0$, the condition

$$mr \in I_n \quad \text{implies} \quad r \in I_n$$

is immediate from the definition of the adic filtration (Example 6.4). □

6.2 Power Series Rings

The purpose of this section is to show that any filtered power series or polynomial ring R as in Corollary 6.12 (with at least one $e_j < \infty$) admits uncountably many mutually non-isomorphic filtered λ-ring structures. In the next two sections, we will describe explicitly all these isomorphism classes of filtered λ-ring structures when R is the truncated polynomial ring $\mathbf{Z}[x]/\langle x^2 \rangle$ or $\mathbf{Z}[x]/\langle x^3 \rangle$.

Here is the main result of this section.

Theorem 6.13. *Consider the filtered ring*

$$R = \frac{A[[x_1, \ldots, x_n]]}{\langle x_1^{e_1}, \ldots, x_n^{e_n} \rangle} \quad or \quad \frac{A[x_1, \ldots, x_n]}{\langle x_1^{e_1}, \ldots, x_n^{e_n} \rangle}$$

with the adic filtration (Example 6.4), for any subring $A \subseteq \mathbf{Q}$, $n \geq 1$, and exponents e_i with $2 \leq e_i \leq \infty$, such that at least one $e_j < \infty$. Then R admits uncountably many isomorphism classes of filtered λ-ring structures.

The plan to prove Theorem 6.13 is to construct some polynomials $\psi^p(x_i)$ satisfying the hypotheses of Corollary 6.12, obtaining some filtered λ-ring structures on R. Different sets of polynomials $\psi^p(x_i)$ are then shown to correspond to non-isomorphic filtered λ-rings.

Proof. [Proof of Theorem 6.13] This proof works equally well for the power series case and the polynomial case. Thus, we will prove them at the same time.

Let $N = \max\{e_j : e_j < \infty\}$, the maximum of those exponents e_j that are finite. For each prime $p \geq N$ and each index j with $e_j < \infty$, choose a positive integer

$$b_{p,j} \in \{p^2, p^3, p^4, \ldots\} \quad \text{with} \quad b_{p,j} \geq e_j. \tag{6.3}$$

There are uncountably many different such sets

$$\begin{aligned}\{b_{p,j}\} &= \{b_{p,j} : \text{prime } p \geq N \text{ and } e_j < \infty\} \\ &= \coprod_{p \geq N} \{b_{p,j} : e_j < \infty\},\end{aligned}$$

where $\coprod$ denotes disjoint union of sets. Indeed, there is at least one e_j that is finite and, for each of the countably infinitely many primes $p \geq N$, there are countably infinitely many choices of $b_{p,j}$. We will use these $b_{p,j}$ to construct the desired Adams operations (and hence filtered λ-ring structures). Moreover, different sets $\{b_{p,j}\}$ correspond to non-isomorphic filtered λ-ring structures on R.

Let $\{b_{p,j}\}$ be a set of integers as in (6.3). For each prime p and $i \in \{1, \ldots, n\}$, consider the following polynomials in R:

$$\begin{aligned}\psi^p(x_i) &= \begin{cases}(1+x_i)^{b_{p,i}} - 1 & \text{if } p \geq N \text{ and } e_i < \infty,\\ (1+x_i)^p - 1 & \text{otherwise,}\end{cases}\\ &= \begin{cases} b_{p,i}x_i + \cdots + x_i^{b_{p,i}} & \text{if } p \geq N \text{ and } e_i < \infty,\\ px_i + \cdots + x_i^p & \text{otherwise.}\end{cases}\end{aligned} \tag{6.4}$$

As in Example 6.4, for each prime p, the polynomials $\psi^p(x_i)$ $(1 \leq i \leq n)$ extend uniquely to a filtered ring homomorphism $\psi^p \colon R \to R$. We first claim that these ψ^p determine a unique filtered λ-ring structure on R.

Since each $\psi^p(x_i)$ has constant term 0, using Corollary 6.12 it remains to show two things:

$$\psi^p(\psi^q(x_i)) = \psi^q(\psi^p(x_i)) \tag{6.5}$$

as polynomials for any primes p and q, and

$$\psi^p(x_i) \equiv x_i^p \pmod{pR}. \tag{6.6}$$

For (6.6) first note that, modulo pR, the p^kth power map $(k \geq 1)$ is linear. Thus, if either $p < N$ or $e_i = \infty$, then we have

$$\psi^p(x_i) = (1+x_i)^p - 1 \equiv x_i^p \pmod{pR}.$$

Likewise, if $p \geq N$ and $e_i < \infty$, then $b_{p,i}$ is a power of p, and we have

$$\begin{aligned}\psi^p(x_i) &= (1+x_i)^{b_{p,i}} - 1\\ &\equiv x_i^{b_{p,i}} \pmod{pR}\\ &\equiv 0 \pmod{pR} \quad (\text{since } b_{p,i} \geq e_i)\\ &\equiv x_i^p \pmod{pR} \quad (\text{since } p \geq N \geq e_i).\end{aligned}$$

This proves (6.6).

For (6.5) suppose that $\psi^p(x_i) = (1+x_i)^b - 1$ and $\psi^q(x_i) = (1+x_i)^c - 1$, where b can be equal to p or of the form (6.3) and similarly for c. Then we have

$$\begin{aligned}\psi^p\left(\psi^q(x_i)\right) &= \left(1+\psi^q(x_i)\right)^b - 1\\ &= \left((1+x_i)^c\right)^b - 1\\ &= (1+x_i)^{cb} - 1\\ &= (1+x_i)^{bc} - 1\\ &= \psi^q\left(\psi^p(x_i)\right).\end{aligned}$$

Therefore, by Corollary 6.12 the ψ^p (6.4) are the Adams operations of a unique filtered λ-ring structure on R.

Let us denote by R_1 the above filtered λ-ring structure on R. Suppose that R_2 is another such filtered λ-ring structure on R given by another set $\{\bar{b}_{p,j}\}$ of integers as in (6.3). Assume that there exists a prime $q \geq N$ such that

$$\{b_{q,j} \colon e_j < \infty\} \neq \{\bar{b}_{q,j} \colon e_j < \infty\} \tag{6.7}$$

as finite sets. We claim that R_1 and R_2 are not isomorphic as filtered λ-rings. This will finish the proof of the Theorem.

Suppose to the contrary that R_1 and R_2 are isomorphic. Thus, there exists a filtered λ-ring isomorphism $\varphi\colon R_1 \to R_2$. Pick any index j with $e_j < \infty$. Since φ is a filtered ring isomorphism, we must have

$$\varphi(x_j) \equiv a_1x_1 + \cdots + a_nx_n \pmod{I_{d+1}}, \tag{6.8}$$

where I_{d+1} is the $(d+1)$st filtration ideal in R and the $a_i \in A$, not all of them equal to 0. (Recall that each variable x_i is assigned a fixed positive filtration d.)

For the following calculation, we set $\bar{b}_{q,i} = q$ if $e_i = \infty$. On the one hand, we have

$$\begin{aligned}\varphi\left(\psi^q(x_j)\right) &\equiv \varphi(b_{q,j}x_j) \pmod{I_{d+1}}\\ &\equiv b_{q,j}(a_1x_1 + \cdots + a_nx_n) \pmod{I_{d+1}}.\end{aligned}$$

On the other hand, we have

$$\begin{aligned}\psi^q\left(\varphi(x_j)\right) &\equiv \psi^q(a_1x_1 + \cdots + a_nx_n) \pmod{I_{d+1}}\\ &\equiv a_1\bar{b}_{q,1}x_1 + \cdots + a_n\bar{b}_{q,n}x_n \pmod{I_{d+1}}\end{aligned}$$

Since there is at least one $a_k \neq 0$, it follows from the equality $\varphi\left(\psi^q(x_j)\right) = \psi^q\left(\varphi(x_j)\right)$ that $b_{q,j}a_k = a_k\bar{b}_{q,k}$. Thus, we have $b_{q,j} = \bar{b}_{q,k}$. Note that this implies that $e_k < \infty$, since $b_{q,j} \neq q$ by construction (6.3). Since j is arbitrary (as long as $e_j < \infty$), it follows that

$$\{b_{q,j}\colon e_j < \infty\} \subseteq \{\bar{b}_{q,j}\colon e_j < \infty\}.$$

By symmetry we conclude that

$$\{b_{q,j}\colon e_j < \infty\} = \{\bar{b}_{q,j}\colon e_j < \infty\}.$$

This is in contradiction with the assumption (6.7). Therefore, R_1 and R_2 are not isomorphic as filtered λ-rings.

This finishes the proof of Theorem 6.13. □

6.3 Classification of $\lambda\left(\mathbf{Z}[x]/\langle x^2\rangle\right)$

The smallest case in Theorem 6.13 is the truncated polynomial ring $R = \mathbf{Z}[x]/\langle x^2\rangle$, also known as the dual number ring. So there are uncountably many mutually non-isomorphic filtered λ-ring structures on the dual number ring. The purpose of this section is to completely classify these isomorphism classes of filtered λ-ring structures (Corollary 6.17).

Definition 6.14. Let R be a (filtered) ring. Denote by $\lambda(R)$ the set of isomorphism classes of (filtered) λ-ring structures on R.

Example 6.15. For the ring $\mathbf{Z}$ of integers, $\lambda(\mathbf{Z})$ is the one-element set because $\mathbf{Z}$ has a unique λ-ring structure (Proposition 1.29 (1)).

The main result of this section makes use of the following result and its proof, which are of independent interests. Recall from Definition 5.1 that a binomial ring is a **Z**-torsion-free ring that is closed under taking binomial symbols. Several characterizations of binomial rings are summarized in Corollary 5.35.

Theorem 6.16. *Let B be a binomial ring. Consider the filtered truncated polynomial ring $B[x]/\langle x^2 \rangle$, in which the variable x has a fixed positive filtration d and elements in B have filtration 0. Then the set $\lambda(B[x]/\langle x^2 \rangle)$ of isomorphism classes of filtered λ-ring structures on $B[x]/\langle x^2 \rangle$ is uncountable.*

Note that $B[x]/\langle x^2 \rangle$ has a two-stage filtration:

$$B[x]/\langle x^2 \rangle = I_0 \supset I_1 = \cdots = I_d = \{bx\colon b \in B\}. \tag{6.9}$$

All the higher filtration ideals are equal to 0.

The proof of Theorem 6.16 is similar to that of Theorem 6.13. We will write down some polynomials and use the filtered version of Wilkerson's Theorem to obtain the desired filtered λ-ring structures.

Proof. [Proof of Theorem 6.16] Let (a_p) be a sequence of elements in B, in which p runs through the primes with

$$a_p \in p\mathbf{Z} \subseteq pB. \tag{6.10}$$

For each prime p, let $\psi^p\colon B[x]/\langle x^2 \rangle \to B[x]/\langle x^2 \rangle$ be the map defined by

$$\psi^p(a + bx) = a + ba_p x \tag{6.11}$$

for $a, b \in B$. It is immediate that ψ^p is a filtered ring homomorphism. (The reader should check this.) We claim that these ψ^p are the Adams operations of a unique filtered λ-ring structure on $B[x]/\langle x^2 \rangle$.

It suffices to check the three conditions in the Filtered Wilkerson's Theorem 6.11. We have

$$\begin{aligned} \psi^p(\psi^q(a + bx)) &= \psi^p(a + ba_q x) \\ &= a + ba_q a_p x \\ &= \psi^q(\psi^p(a + bx)). \end{aligned}$$

Thus, for any primes p and q, the maps ψ^p and ψ^q commute. For the second condition in Theorem 6.11, we have

$$\begin{aligned} \psi^p(a + bx) &= a + ba_p x \\ &\equiv a \quad (\text{mod } pB[x]/\langle x^2 \rangle) \text{ by (6.10)} \\ &\equiv a^p \quad (\text{mod } pB[x]/\langle x^2 \rangle) \text{ by Lemma 5.5} \\ &\equiv a^p + b^p x^p \quad (\text{mod } pB[x]/\langle x^2 \rangle) \text{ since } x^p = 0 \\ &\equiv (a + bx)^p \quad (\text{mod } pB[x]/\langle x^2 \rangle). \end{aligned}$$

Using (6.9) the last condition in Theorem 6.11 is clear, since B is **Z**-torsion-free by assumption. Therefore, by Theorem 6.11 the ψ^p (6.11) are the Adams operations of a unique λ-ring structure on $B[x]/\langle x^2 \rangle$.

Denote by R_1 the filtered λ-ring structure on $B[x]/\langle x^2 \rangle$ constructed above. Let R_2 be another such filtered λ-ring structure corresponding to a sequence (b_p) of elements in B as in (6.10). Assume that R_1 and R_2 are isomorphic filtered λ-rings. We claim that $a_p = b_p$ for each prime p. Indeed, suppose that $\varphi\colon R_1 \xrightarrow{\cong} R_2$ is a filtered λ-isomorphism. Then we must have $\varphi(x) = ux$ for some invertible element $u \in B$. On the one hand, we have

$$\varphi(\psi^p(x)) = \varphi(a_p x) = a_p ux,$$

since a_p is an integer. On the other hand, we have

$$\psi^p(\varphi(x)) = \psi^p(ux) = ub_p x$$

by the definition of ψ^p (6.11) in R_2. Since φ is a λ-homomorphism, it is compatible with the Adams operations as well (Theorem 3.6). Thus, we have $a_p u = ub_p$, which implies that $a_p = b_p$ because u is invertible.

Finally, since B is **Z**-torsion-free, there are uncountably many distinct sequences (a_p) satisfying (6.10). Each such sequence determines uniquely a filtered λ-ring structure on $B[x]/\langle x^2 \rangle$. Moreover, the previous paragraph shows that the filtered λ-ring structures on $B[x]/\langle x^2 \rangle$ corresponding to distinct sequences (a_p) are not isomorphic. This finishes the proof of Theorem 6.16. □

Next we observe that, when $B = \mathbf{Z}$, every filtered λ-ring structure on $\mathbf{Z}[x]/\langle x^2 \rangle$ is, in fact, one of those described in Theorem 6.16. Recall that $\lambda(\mathbf{Z}[x]/\langle x^2 \rangle)$ denotes the set of isomorphism classes of filtered λ-ring structures on $\mathbf{Z}[x]/\langle x^2 \rangle$.

Corollary 6.17. *There is a bijection*

$$\alpha\colon \prod_{p\ prime} p\mathbf{Z} \xrightarrow{\cong} \lambda(\mathbf{Z}[x]/\langle x^2 \rangle)$$

that sends a sequence $(a_p) \in \prod_p p\mathbf{Z}$ to the (isomorphism class of the) filtered λ-ring structure on $\mathbf{Z}[x]/\langle x^2 \rangle$ with Adams operations

$$\psi^p(a + bx) = a + ba_p x$$

as in (6.11).

Proof. By the proof of Theorem 6.16 the map α is injective. To see that α is surjective, it suffices to show that every filtered λ-ring structure on $\mathbf{Z}[x]/\langle x^2 \rangle$ must have Adams operations of the form $\psi^p(a+bx) = a+ba_p x$ for some integers $a_p \in p\mathbf{Z}$ for p prime.

So fix a filtered λ-ring structure on $\mathbf{Z}[x]/\langle x^2 \rangle$. Since its Adams operations are filtered ring homomorphisms (Proposition 6.10), for each prime p we have $\psi^p(x) = a_p x$ for some integer a_p. Thus, given integers a and b, we have $\psi^p(a+bx) = a+ba_p x$. Moreover, we have

$$\begin{aligned} a_p x &= \psi^p(x) \\ &\equiv x^p \pmod{p\mathbf{Z}[x]/\langle x^2 \rangle} \quad \text{by (3.4)} \\ &\equiv 0 \pmod{p\mathbf{Z}[x]/\langle x^2 \rangle}. \end{aligned}$$

This implies that $a_p \in p\mathbf{Z}$, as desired. □

In the following example, we describe the λ-operations on the uncountably many mutually non-isomorphic filtered λ-ring structures on $\mathbf{Z}[x]/\langle x^2\rangle$.

Example 6.18. Suppose that a filtered λ-ring structure on $\mathbf{Z}[x]/\langle x^2\rangle$ has Adams operations $\psi^p(x) = a_p x$ for some integers $a_p \in p\mathbf{Z}$ for p primes. Define the integers a_n for $n \geq 1$ by setting

$$a_n = \begin{cases} 1 & \text{if } n = 1, \\ a_{p_1}^{e_1} \cdots a_{p_r}^{e_r} & \text{if } n \text{ has prime factorization } p_1^{e_1} \cdots p_r^{e_r}. \end{cases}$$

Then we claim that the λ-operations on x are given by

$$\lambda^n(x) = (-1)^{n-1} \frac{a_n}{n} x \tag{6.12}$$

for $n \geq 1$. Note that a_n is divisible by n, since each a_p is divisible by p.

To prove (6.12), first observe that

$$\psi^n(x) = (\psi^{p_1})^{e_1} \cdots (\psi^{p_r})^{e_r}(x) = a_n x.$$

From the definition (3.1) of Adams operations, we have

$$\sum_{n\geq 1} (-1)^n (a_n x) t^n = \psi_{-t}(x) = -t \frac{d}{dt} (\log \lambda_t(x)).$$

Dividing by $-t$ and integrating with respect to t, we obtain

$$\log \lambda_t(x) = \sum_{n\geq 1} (-1)^{n-1} \frac{a_n x}{n} t^n.$$

Exponentiating both sides and using $x^2 = 0$, we have

$$\begin{aligned}
\lambda_t(x) &= \exp\left(\sum_{n\geq 1} (-1)^{n-1} \frac{a_n x}{n} t^n \right) \\
&= \prod_{n\geq 1} \exp\left((-1)^{n-1} \frac{a_n x}{n} t^n \right) \\
&= \prod_{n\geq 1} \left(1 + (-1)^{n-1} \frac{a_n x}{n} t^n \right) \\
&= 1 + a_1 x t - \frac{a_2 x}{2} t^2 + \frac{a_3 x}{3} t^3 - \cdots.
\end{aligned}$$

Equating the coefficients of t^n on both sides, we obtain the expression (6.12) for $\lambda^n(x)$.

A general element in $\mathbf{Z}[x]/\langle x^2\rangle$ has the form $a + bx$ for some integers a and b. Thus, using $x^2 = 0$ we have

$$\begin{aligned}
\lambda_t(a + bx) &= \lambda_t(a)\lambda_t(x)^b \quad \text{by (1.8) and Proposition 1.13 (4)} \\
&= (1+t)^a \lambda_t(x)^b \quad \text{by Proposition 1.29 (1)} \\
&= (1+t)^a \left(1 + \sum_{n\geq 1} (-1)^{n-1} \frac{a_n x}{n} t^n \right)^b \\
&= (1+t)^a \left(1 + \sum_{n\geq 1} (-1)^{n-1} \frac{b a_n x}{n} t^n \right).
\end{aligned}$$

From this expression one can determine any $\lambda^m(a + bx)$.

6.4 Classification of $\lambda\left(\mathbf{Z}[x]/\langle x^3\rangle\right)$

The purpose of this section is to describe explicitly all of the uncountably many isomorphism classes of filtered λ-ring structures (Theorem 6.13) on the truncated polynomial ring $\mathbf{Z}[x]/\langle x^3\rangle$. This classification is presented in Theorems 6.19 and 6.21. As in the previous sections, the variable x is given a fixed positive filtration d, and elements in $\mathbf{Z}$ have filtration 0.

From Corollary 6.17 we know that the set $\lambda(\mathbf{Z}[x]/\langle x^2\rangle)$ of isomorphism classes of filtered λ-ring structures on $\mathbf{Z}[x]/\langle x^2\rangle$ admits a rather simple description. Namely, it is in bijection with the countable product $\prod_p p\mathbf{Z}$, where p runs through the primes. The case of $\mathbf{Z}[x]/\langle x^3\rangle$, on the other hand, is more complicated.

Using Corollary 6.12 we will describe any filtered λ-ring structure on $\mathbf{Z}[x]/\langle x^3\rangle$ in terms of its Adams operations ψ^p (p prime) when applied to the variable x. Since the polynomial $\psi^p(x) \in \mathbf{Z}[x]/\langle x^3\rangle$ must have constant term 0 (Corollary 6.12), it has the form

$$\psi^p(x) = b_p x + c_p x^2$$

for some integers b_p and c_p. The conditions

$$\begin{aligned}\psi^p(\psi^q(x)) &= \psi^q(\psi^p(x)),\\ \psi^p(x) &\equiv x^p \pmod{p\mathbf{Z}[x]/\langle x^3\rangle}\end{aligned}$$

impose certain conditions on the coefficients b_p and c_p. There are two cases: $b_2 = 0$ or $b_2 \neq 0$. The first case is relatively easy to deal with and is contained in the following result. Proofs will be given at the end of this section.

Theorem 6.19. *Let R be a filtered λ-ring structure on $\mathbf{Z}[x]/\langle x^3\rangle$ with Adams operations ψ_R^n ($n \geq 1$). Suppose that*

$$\psi_R^2(x) = mx^2 \tag{6.13}$$

for some integer m (i.e., $\psi_R^2(x)$ has zero linear term). Then R is isomorphic to one of the filtered λ-rings

$$S((c_p)) = \{\psi^p(x) = c_p x^2 : p \text{ prime}\} \tag{6.14}$$

with

$$c_2 \equiv 1 \pmod 2 \quad \text{and} \quad c_p \equiv 0 \pmod p$$

for odd primes p. Conversely, any such sequence (c_p) determines uniquely a filtered λ-ring structure $S((c_p))$ on $\mathbf{Z}[x]/\langle x^3\rangle$. Moreover, two such filtered λ-ring structures $S((c_p))$ and $S((c'_p))$ are isomorphic if and only if there exists $u \in \{\pm 1\}$ such that $c_p = uc'_p$ for each prime p.

In other words, the subset of $\lambda(\mathbf{Z}[x]/\langle x^3\rangle)$ consisting of the isomorphism classes of filtered λ-ring structures on $\mathbf{Z}[x]/\langle x^3\rangle$ that satisfy (6.13) is in bijection with the quotient set $\left((1+2\mathbf{Z}) \times \prod_{p>2} p\mathbf{Z}\right)/\pm$. It is clear that this is an uncountable set. At the end of this section, we will compute the λ-operations in $S((c_p))$.

On the other hand, suppose that $b_2 \neq 0$. We need the following notation.

Definition 6.20. Let p be a prime and n be an integer. Define $\nu_p(n)$ by

$$\nu_p(n) = \begin{cases} \infty & \text{if } n = 0, \\ \text{largest integer such that } p^{\nu_p(n)} \text{ divides } n & \text{if } n \neq 0. \end{cases}$$

The following two sets of technical conditions are needed in order to describe the relationships between the various coefficients b_p and c_p.

$$\begin{cases} b_2 \neq 0, \\ b_p \equiv 0 \pmod{p} \text{ for all primes } p, \text{ and} \\ b_p(b_p - 1) \equiv 0 \pmod{2^{\nu_2(b_2)}} \text{ for all odd primes } p. \end{cases} \tag{6.15}$$

Let (b_p) be integers, one for each prime p, satisfying (6.15). Consider an odd prime p (if there exists any) such that

$$\begin{cases} b_p \neq 0, \text{ and} \\ \nu_p(b_p) = \min\{\, \nu_p(b_q(b_q - 1)) : q \text{ prime}\,\}. \end{cases} \tag{6.16}$$

In (6.16), since p divides b_p, it does not divide $(b_p - 1)$. Thus, we have

$$\begin{aligned} \nu_p(b_p) &= \min\{\, \nu_p(b_q(b_q - 1)) : q \text{ prime}\,\} \\ &= \nu_p(b_p(b_p - 1)). \end{aligned}$$

Moreover, since $b_2 \neq 0, 1$, it follows that if p satisfies (6.16), then $p \mid b_2(b_2 - 1) \neq 0$. In particular, there are at most finitely many odd primes p that satisfy (6.16).

If (b_p) satisfies (6.15), then we write

$$G = \gcd(b_p(b_p - 1)\colon \text{all primes } p), \tag{6.17}$$

the greatest common divisor of $b_p(b_p-1)$ for p prime. The following result completes the classification of the set $\lambda(\mathbf{Z}[x]/\langle x^3\rangle)$ of isomorphism classes of filtered λ-ring structures on $\mathbf{Z}[x]/\langle x^3\rangle$.

Theorem 6.21. *Let R be a filtered λ-ring structure on $\mathbf{Z}[x]/\langle x^3\rangle$ with Adams operations ψ_R^n $(n \geq 1)$. Suppose that*

$$\psi_R^2(x) = mx + nx^2 \tag{6.18}$$

for some non-zero integer m. Then R is isomorphic to one of the filtered λ-rings

$$S((b_p), k) = \left\{ \psi^p(x) = b_p x + \frac{k b_p(b_p - 1)}{G} x^2 : p \text{ prime} \right\} \tag{6.19}$$

with (b_p) satisfying (6.15) and k satisfying the following conditions. Let $p_1, \ldots, p_n$ be the list of all the distinct odd primes satisfying (6.16). Then k is an odd integer such that

$$\begin{cases} 1 \leq k \leq \dfrac{G}{2}, \\ k \equiv 0 \pmod{p_1 \cdots p_n}. \end{cases} \tag{6.20}$$

Conversely, any such pair $((b_p), k)$ gives rise to a filtered λ-ring structure $S((b_p), k)$ on $\mathbf{Z}[x]/\langle x^3\rangle$. Two such filtered λ-rings $S((b_p), k)$ and $S((b'_p), k')$ are isomorphic if and only if $((b_p), k) = ((b'_p), k')$. Moreover, no $S((b_p), k)$ is isomorphic to any $S((c_p))$ (6.14).

In Theorem 6.21 if there is no odd prime satisfying (6.16), then $p_1 \cdots p_n$ is the empty product, i.e., 1. In this case, the second condition in (6.20) is trivially true and imposes no additional restriction on k.

Theorems 6.19 and 6.21 together give the complete classification of the set $\lambda(\mathbf{Z}[x]/\langle x^3\rangle)$ of filtered λ-ring structures on $\mathbf{Z}[x]/\langle x^3\rangle$. Note that there are uncountably many distinct pairs $((b_p), k)$ with (b_p) and k satisfying (6.15) and (6.20), respectively. In particular, there are uncountably many mutually non-isomorphic filtered λ-ring structures on $\mathbf{Z}[x]/\langle x^3\rangle$ that satisfy (6.18).

On the other hand, given (b_p) satisfying (6.15), there are only finitely many odd integers k that satisfy (6.20). The exact number of such k is given in the following result. For a real number s, let $\lceil s\rceil$ denote the smallest integer that is greater than or equal to s.

Corollary 6.22. *Let (b_p) be a fixed sequence of integers satisfying* (6.15). *Then there are exactly*

$$\left\lceil \frac{G}{4p_1\cdots p_n} \right\rceil = \left\lceil \frac{\gcd\left(b_p(b_p-1)\colon \text{all primes } p\right)}{4p_1\cdots p_n} \right\rceil \tag{6.21}$$

filtered λ-rings of the form $S((b_p), k)$ (6.19).

Proof. With the notations of Theorem 6.21, each prime p_i is odd. Thus, k is a positive odd multiple of the product $p_1 \cdots p_n$, and it is at most $G/2$. The exact number of such k is stated in (6.21). □

Example 6.23. Suppose that $b_p = p$ for each prime p. Then the sequence $(b_p) = (p)$ satisfies (6.15). In this case, we have

$$G = \gcd\left(p(p-1)\colon \text{all primes } p\right) = 2.$$

This implies that $k = 1$ is the only odd integer that satisfies (6.20). Thus, the only filtered λ-ring structure on $\mathbf{Z}[x]/\langle x^3\rangle$ of the form $S((p), k)$ is

$$S((p), 1) = \left\{ \psi^p(x) = px + \frac{p(p-1)}{2}x^2 \colon p \text{ prime} \right\}.$$

In other words, its Adams operation is $\psi^p(x) = (1+x)^p - 1$ for each prime p. Moreover, let $\mathbf{CP}^2$ denote the complex projective 2-space, and let $K(\mathbf{CP}^2)$ denote its K-theory. Using some homotopy theory, it can be shown that there is an isomorphism $K(\mathbf{CP}^2) \cong S((p), 1)$ of filtered λ-rings with x in filtration 2.

Example 6.24. Suppose that $b_p = p^2$ for each prime p. Then the sequence $(b_p) = (p^2)$ satisfies (6.15). There are no odd primes that satisfy (6.16), and

$$G = \gcd\left(p^2(p^2-1)\colon \text{all primes } p\right) = 12.$$

This implies that there are three odd integers k that satisfy (6.20): $k = 1$, 3, and 5. Each such $k \in \{1, 3, 5\}$ corresponds to a filtered λ-ring structure on $\mathbf{Z}[x]/\langle x^3 \rangle$ of the form

$$S((p^2), k) = \left\{ \psi^p(x) = p^2 x + \frac{kp^2(p^2-1)}{12} x^2 : p \text{ prime} \right\}.$$

Moreover, let $\mathbf{HP}^2$ denote the quaternionic projective 2-space, and denote its K-theory by $K(\mathbf{HP}^2)$. Using some homotopy theory, it can be shown that there is an isomorphism $K(\mathbf{HP}^2) \cong S((p^2), 1)$ of filtered λ-rings with x in filtration 4.

We now begin to prove Theorems 6.19 and 6.21 with a series of Lemmas. The steps are as follows:

(1) Establish conditions on the coefficients of the polynomials

$$\psi^p(x) = b_p x + c_p x^2$$

under which the ψ^p are the Adams operations on a filtered λ-ring structure on $\mathbf{Z}[x]/\langle x^3 \rangle$.

(2) Show that $S((c_p))$ (6.14) and $S((b_p), k)$ (6.19) are, in fact, filtered λ-ring structures on $\mathbf{Z}[x]/\langle x^3 \rangle$.

(3) Establish criteria under which two filtered λ-ring structures on $\mathbf{Z}[x]/\langle x^3 \rangle$ are isomorphic.

(4) Show that $S((c_p))$ and $S((b_p), k)$ are all non-isomorphic to each other, except when $(c_p) = \pm(c'_p)$.

(5) Show that every filtered λ-ring structure on $\mathbf{Z}[x]/\langle x^3 \rangle$ is isomorphic to one of the $S((c_p))$ or $S((b_p), k)$.

The first preliminary result is a version of Corollary 6.12 for the truncated polynomial ring $\mathbf{Z}[x]/\langle x^3 \rangle$.

Lemma 6.25. *Consider the polynomials*

$$\{\psi^p(x) = b_p x + c_p x^2 : p \textit{ prime}\}$$

in $\mathbf{Z}[x]/\langle x^3 \rangle$. Then they are the Adams operations of a filtered λ-ring structure on $\mathbf{Z}[x]/\langle x^3 \rangle$ if and only if the following four conditions hold:

$$\begin{aligned} b_p &\equiv 0 \pmod{p} \quad \textit{for all primes } p, \\ c_2 &\equiv 1 \pmod{2}, \\ c_p &\equiv 0 \pmod{p} \quad \textit{for all odd primes } p, \\ (b_p^2 - b_p)c_q &= (b_q^2 - b_q)c_p \quad \textit{for all primes } p \textit{ and } q. \end{aligned} \tag{6.22}$$

Proof. By Corollary 6.12 the given collection $\psi^p(x)$ extends to a unique filtered λ-ring structure on $\mathbf{Z}[x]/\langle x^3 \rangle$ if and only if

$$\psi^p(\psi^q(x)) = \psi^q(\psi^p(x)) \tag{6.23}$$

and

$$\psi^p(x) \equiv x^p \pmod{p\mathbf{Z}[x]/\langle x^3\rangle} \tag{6.24}$$

for all primes p and q. We have

$$\begin{aligned}\psi^p(\psi^q(x)) &= b_p(b_qx + c_qx^2) + c_p(b_qx + c_qx^2)^2 \\ &= (b_pb_q)x + (b_pc_q + b_q^2c_p)x^2.\end{aligned}$$

Likewise, we have

$$\psi^q(\psi^p(x)) = (b_qb_p)x + (b_qc_p + b_p^2c_q)x^2.$$

Thus, (6.23) is equivalent to the condition

$$b_pc_q + b_q^2c_p = b_qc_p + b_p^2c_q,$$

which in turn is equivalent to the last condition in (6.22). Similarly, since $x^3 = 0$, the condition (6.24) is equivalent to the first three conditions in (6.22). □

We use this Lemma to observe that $S((c_p))$ and $S((b_p), k)$ are indeed filtered λ-ring structures on $\mathbf{Z}[x]/\langle x^3\rangle$.

Lemma 6.26. *There exist filtered λ-ring structures $S((c_p))$ and $S((b_p), k)$ on $\mathbf{Z}[x]/\langle x^3\rangle$ as stated in* (6.14) *and* (6.19), *respectively.*

Proof. Using Lemma 6.25 it suffices to show that the four conditions in (6.22) are satisfied. For $S((c_p)) = \{\psi^p(x) = c_px^2 : p \text{ prime}\}$ with $c_2 \equiv 1 \pmod 2$ and $c_p \equiv 0 \pmod p$, the conditions in (6.22) are clearly satisfied, since each $b_p = 0$.

Next suppose that we are given

$$S((b_p), k) = \left\{\psi^p(x) = b_px + \frac{kb_p(b_p-1)}{G}x^2 : p \text{ prime}\right\}$$

as in (6.19) with (b_p) satisfying (6.15) and k satisfying (6.20). We already know that $b_p \equiv 0 \pmod p$ for each prime p. For the last condition in (6.22), first note that $c_p = (kb_p(b_p-1))/G$. It follows that

$$(b_p^2 - b_p)c_q = \frac{k(b_q^2 - b_q)(b_p^2 - b_p)}{G} = (b_q^2 - b_q)c_p.$$

To show that $c_2 \equiv 1 \pmod 2$, note that $2^{\nu_2(b_2)} \mid b_p(b_p-1)$ for every prime p. It follows that $2^{\nu_2(b_2)} \mid G$. But any power of 2 that divides G must also divide b_2. Thus, we have $\nu_2(b_2) = \nu_2(G)$. Since both k and $(b_2 - 1)$ are odd, it follows that $c_2 = (kb_2(b_2-1))/G$ is odd.

Finally, to show that $c_p \equiv 0 \pmod p$, first suppose that p satisfies (6.16). In this case, we have $k \equiv 0 \pmod p$. This implies that

$$c_p = k\frac{b_p(b_p-1)}{G} \equiv 0 \pmod p,$$

as desired.

Next suppose that p does not satisfy (6.16). If $b_p = 0$, then $c_p = 0$, which is divisible by p. On the other hand, suppose that $b_p \neq 0$. Then we have

$$\begin{aligned}\nu_p(b_p) &= \nu_p(b_p(b_p - 1)) \\ &> \min\{\, \nu_p\,(b_q(b_q - 1)) : q \text{ prime}\,\} \\ &= \nu_p(G).\end{aligned}$$

This implies that

$$\nu_p(c_p) = \nu_p\left(\frac{kb_p(b_p - 1)}{G}\right) > 1,$$

and c_p is divisible by p, as desired. □

Since ψ^2 commutes with ψ^p, the coefficient b_2 imposes strong restrictions on the coefficients b_p for odd primes p.

Lemma 6.27. *Suppose that b_p and c_p (p prime) are integers that satisfy* (6.22) *in Lemma 6.25.*

(1) If $b_2 = 0$, then $b_p = 0$ for all odd primes p.
(2) If $b_2 \neq 0$, then $b_p^2 - b_p \equiv 0 \pmod{2^{\nu_2(b_2)}}$ for all odd primes p.

Proof. First suppose that $b_2 = 0$. Then the last condition in (6.22) (with p odd and $q = 2$) becomes $(b_p^2 - b_p)c_2 = 0$. Since c_2 is odd, it is not equal to zero. Thus, we have $b_p^2 - b_p = 0$. Since b_p is divisible by p, we have $b_p \neq 1$, which implies that $b_p = 0$, as desired.

Next suppose that $b_2 \neq 0$. From the last condition in (6.22) we have $(b_p^2 - b_p)c_2 = (b_2^2 - b_2)c_p$ with p odd, it follows that $(b_p^2 - b_p)c_2 \equiv 0 \pmod{2^{\nu_2(b_2)}}$. Since c_2 is odd, it follows that $b_p^2 - b_p \equiv 0 \pmod{2^{\nu_2(b_2)}}$, as desired. □

The following result provides necessary and sufficient conditions under which two filtered λ-ring structures on $\mathbf{Z}[x]/\langle x^3\rangle$ are isomorphic.

Lemma 6.28. *Let*

$$S = \{\psi^p(x) = b_p x + c_p x^2\} \quad \textit{and} \quad \overline{S} = \{\psi^p(x) = \overline{b}_p x + \overline{c}_p x^2\}$$

be two filtered λ-ring structures on $\mathbf{Z}[x]/\langle x^3\rangle$, where p runs through the primes. Then S is isomorphic to $\overline{S}$ as filtered λ-rings if and only if the following two conditions are satisfied:

(1) $b_p = \overline{b}_p$ for each prime p.
(2) (a) If $b_2 = \overline{b}_2 = 0$, then there exists $u \in \{\pm 1\}$ such that $c_p = u\overline{c}_p$ for each prime p.
(b) If $b_2 = \overline{b}_2 \neq 0$, then there exist $u \in \{\pm 1\}$ and $a \in \mathbf{Z}$ such that

$$ab_2(b_2 - 1) = c_2 - u\overline{c}_2. \tag{6.25}$$

Proof. For the "only if" part, suppose that S and $\overline{S}$ are isomorphic. So there exists a filtered λ-isomorphism $\varphi\colon S \xrightarrow{\cong} \overline{S}$, and $\varphi(x) = ux + ax^2$ for some $u \in \{\pm 1\}$ and $a \in \mathbf{Z}$. For any prime p, we have

$$\begin{aligned}\varphi\left(\psi^p(x)\right) &= b_p(ux + ax^2) + c_p(ux + ax^2)^2 \\ &= ub_p x + \left(ab_p + c_p\right) x^2.\end{aligned}$$

Likewise, we have

$$\psi^p\left(\varphi(x)\right) = u\overline{b}_p x + \left(a\overline{b}_p^2 + u\overline{c}_p\right) x^2.$$

Equating the coefficients we obtain

$$\begin{aligned}ub_p &= u\overline{b}_p, \\ ab_p + c_p &= a\overline{b}_p^2 + u\overline{c}_p.\end{aligned} \tag{6.26}$$

This first condition in (6.26) implies that $b_p = \overline{b}_p$ because $u \in \{\pm 1\}$. Now if $b_2 = \overline{b}_2 = 0$, then $b_p = \overline{b}_p = 0$ for all primes p by Lemma 6.27. Thus, the second condition in (6.26) implies that $c_p = u\overline{c}_p$. On the other hand, suppose that $b_2 = \overline{b}_2 \neq 0$. Then the second condition in (6.26) with $p = 2$ becomes $c_2 - u\overline{c}_2 = ab_2^2 - ab_2$, which is equivalent to (6.25).

For the "if" part, first suppose that (1) and (2)(a) hold. As above, since $b_2 = 0$, it follows that $b_p = 0$ for all primes p. Using the $u \in \{\pm 1\}$ from condition (2)(a), define a filtered ring isomorphism $\phi\colon S \to \overline{S}$ by setting $\phi(x) = ux$. The same calculation as in the first paragraph of this proof shows that

$$\begin{aligned}\phi(\psi^p(x)) &= c_p x^2 \\ &= u\overline{c}_p x^2 \\ &= \psi^p(\phi(x)).\end{aligned}$$

This implies that ϕ commutes with ψ^p for every prime p, which in turn implies that ϕ commutes with all the ψ^n $(n \geq 1)$. Since $\mathbf{Z}[x]/\langle x^3\rangle$ is $\mathbf{Z}$-torsion-free, it follows from Corollary 3.16 that ϕ is a λ-homomorphism. Thus, we conclude that ϕ is a filtered λ-isomorphism.

Finally, suppose that (1) and (2)(b) hold. Using the $u \in \{\pm 1\}$ and $a \in \mathbf{Z}$ from (2)(b), we have a filtered ring isomorphism $\phi\colon S \to \overline{S}$ defined by $\phi(x) = ux + ax^2$. Its inverse is defined by $\phi^{-1}(x) = ux - uax^2$. Pick a prime p. We claim that ϕ commutes with ψ^p. If $b_p = \overline{b}_p = 0$, then the last condition in (6.22) with $q = 2$ implies that $c_p = 0 = \overline{c}_p$, since $b_2 = \overline{b}_2 \neq 0, 1$. In other words, $\psi^p = 0$ in both S and $\overline{S}$, which of course implies that ϕ commutes with ψ^p.

On the other hand, suppose that $b_p = \overline{b}_p \neq 0$. Note that $b_p = \overline{b}_p \neq 1$ either because b_p is divisible by p. Using (2)(b) and the last condition in (6.22) with $q = 2$, we have

$$\begin{aligned}a &= \frac{c_2 - u\overline{c}_2}{b_2^2 - b_2} = \frac{c_p}{b_p^2 - b_p} - u\frac{\overline{c}_p}{b_p^2 - b_p} \\ &= \frac{c_p - u\overline{c}_p}{b_p^2 - b_p}.\end{aligned}$$

Equivalently, we have $a(b_p^2 - b_p) = c_p - u\bar{c}_p$. Both conditions in (6.26) are now satisfied. Thus, the same calculation as in the first paragraph of this proof shows that ϕ commutes with ψ^p, as desired.

Since $\mathbf{Z}[x]/\langle x^3\rangle$ is $\mathbf{Z}$-torsion-free, it follows from Corollary 3.16 that ϕ is a λ-homomorphism. Thus, we conclude that ϕ is a filtered λ-isomorphism. □

Using the criteria from the previous Lemma, we can now show that the filtered λ-rings $S((c_p))$ and $S((b_p), k)$ are mutually non-isomorphic, except when $(c_p) = \pm(c'_p)$.

Lemma 6.29. *The following statements hold.*

1. *Two filtered λ-rings $S((c_p))$ and $S((c'_p))$ as in* (6.14) *are isomorphic if and only if there exists $u \in \{\pm 1\}$ such that $c_p = uc'_p$ for each prime p.*
2. *Two filtered λ-rings $S((b_p), k)$ and $S((b'_p), k')$ as in* (6.19) *are isomorphic if and only if $((b_p), k) = ((b'_p), k')$.*
3. *No $S((b_p), k)$ is isomorphic to any $S((c_p))$.*

Proof. The first statement is directly contained in Lemma 6.28.

For the second statement, the "if" part is clear. Conversely, suppose that $S((b_p), k)$ and $S((b'_p), k')$ are isomorphic filtered λ-rings. Then $b_p = b'_p$ for each prime p by Lemma 6.28. It remains to show that k and k' are equal.

First note that by (6.20), both k and k' are in the range $1 \le k,\, k' \le G/2$. If both k and k' are equal to $G/2$, then we are done. So suppose that at least one of k and k' is less than $G/2$. By Lemma 6.28 there exist $u \in \{\pm 1\}$ and an integer a such that

$$\begin{aligned} ab_2(b_2 - 1) &= c_2 - uc'_2 \\ &= \frac{kb_2(b_2-1)}{G} - u\frac{k'b_2(b_2-1)}{G}. \end{aligned}$$

Since $b_2 \neq 0$ or 1, this implies that $k - uk' = aG$. Now the absolute value

$$|aG| = |k - uk'| \le k + k' < G.$$

Thus, we must have $a = 0$ and $u = 1$, since both k and k' are positive. This shows that k is equal to k', as desired.

The last assertion again follows from (1) in Lemma 6.28 because $b_2 = 0$ in $S((c_p))$. □

Next we show that the $S((c_p))$ and $S((b_p), k)$ account for all the isomorphism classes of filtered λ-ring structures on $\mathbf{Z}[x]/\langle x^3\rangle$.

Lemma 6.30. *Let R be a filtered λ-ring structure on $\mathbf{Z}[x]/\langle x^3\rangle$ with Adams operations*

$$\psi_R^p(x) = b_p x + c_p x^2$$

for p prime. If $b_2 = 0$, then R is isomorphic to the filtered λ-ring $S((c_p))$ (Lemma 6.26).

Proof. Since $b_2 = 0$, by Lemma 6.27 we have $b_p = 0$ for all primes p. Moreover, we have $c_2 \equiv 1 \pmod 2$ and $c_p \equiv 0 \pmod p$ for odd primes p by Lemma 6.25. Therefore, R is actually equal (and hence isomorphic) to $S((c_p))$. □

Lemma 6.31. *Let R be a filtered λ-ring structure on $\mathbf{Z}[x]/\langle x^3\rangle$ with Adams operations*

$$\psi_R^p(x) = b_p x + c_p x^2$$

for p prime. If $b_2 \neq 0$, then R is isomorphic to the filtered λ-ring $S((b_p), k)$ for some integer k (Lemma 6.26).

Proof. The reader should be aware that this proof, although not difficult, is a bit long.

First consider the case when $b_p = 0$ for all odd primes p. Then the last condition in (6.22) (with $q = 2$) implies that $c_p = 0$ as well for all odd primes p. Since c_2 is odd and b_2 is even, there exists a unique odd integer k in the range $1 \leq k \leq b_2(b_2-1)/2 = G/2$ such that $c_2 \equiv uk \pmod{b_2(b_2-1)}$ for some $u \in \{\pm 1\}$. So there exists an integer a such that

$$ab_2(b_2-1) = c_2 - uk = c_2 - u\frac{kb_2(b_2-1)}{G}.$$

Thus, it follows from Lemma 6.28 that R is isomorphic to $S((b_p), k)$.

Next assume that $b_p \neq 0$ for at least one odd prime p. We can write

$$c_2 = mb_2(b_2-1) + r \tag{6.27}$$

for some integer m and remainder r with $0 < r < b_2(b_2-1)$. The remainder r must be odd, since c_2 is odd and b_2 is even. Define $\overline{c}_p$ for p prime as follows:

$$\begin{aligned}
\overline{c}_2 &= \begin{cases} r & \text{if } 1 \leq r \leq \dfrac{b_2(b_2-1)}{2}, \\ b_2(b_2-1) - r & \text{if } r > \dfrac{b_2(b_2-1)}{2}, \end{cases} \\
\overline{c}_p &= \begin{cases} c_p - mb_p(b_p-1) & \text{if } 1 \leq r \leq \dfrac{b_2(b_2-1)}{2}, \\ (1+m)b_p(b_p-1) - c_p & \text{if } r > \dfrac{b_2(b_2-1)}{2} \end{cases}
\end{aligned} \tag{6.28}$$

for odd primes p. We claim that

$$S = \{\psi^p(x) = b_p x + \overline{c}_p x^2 : p \text{ prime}\} \tag{6.29}$$

is a filtered λ-ring structure on $\mathbf{Z}[x]/\langle x^3\rangle$. By Lemma 6.25 it suffices to prove the four conditions in (6.22). The first condition on b_p is true because R is a filtered λ-ring structure on $\mathbf{Z}[x]/\langle x^3\rangle$. Next, $\overline{c}_2$ is odd because r is odd and b_2 is even. For an odd prime p, since both b_p and c_p are divisible by p, so is $\overline{c}_p$.

For the last condition in (6.22), it suffices to show that

$$\frac{\overline{c}_p}{b_p^2 - b_p} = \frac{\overline{c}_2}{b_2^2 - b_2} \tag{6.30}$$

whenever $b_p \neq 0$ and p is odd. To prove (6.30), first suppose that $r \leq b_2(b_2-1)/2$. In this case, we have

$$\frac{\overline{c}_2}{b_2^2 - b_2} = \frac{r}{b_2^2 - b_2} = \frac{c_2}{b_2^2 - b_2} - m$$

by (6.27). For an odd prime p with $b_p \neq 0$, we have

$$\begin{aligned}\frac{\overline{c}_p}{b_p^2 - b_p} &= \frac{c_p - m(b_p^2 - b_p)}{b_p^2 - b_p} = \frac{c_p}{b_p^2 - b_p} - m \\ &= \frac{c_2}{b_2^2 - b_2} - m,\end{aligned}$$

since the various b_p and c_p satisfy the last condition in (6.22). This proves (6.30) when $r \leq b_2(b_2-1)/2$.

Next suppose that $r > b_2(b_2-1)/2$. In this case, we have

$$\begin{aligned}\frac{\overline{c}_2}{b_2^2 - b_2} &= \frac{b_2^2 - b_2 - r}{b_2^2 - b_2} = 1 - \frac{r}{b_2^2 - b_2} \\ &= 1 - \left(\frac{c_2}{b_2^2 - b_2} - m\right) = 1 + m - \frac{c_2}{b_2^2 - b_2}.\end{aligned}$$

On the other hand, if $b_p \neq 0$ with p odd, then

$$\begin{aligned}\frac{\overline{c}_p}{b_p^2 - b_p} &= \frac{(1+m)(b_p^2 - b_p) - c_p}{b_p^2 - b_p} = 1 + m - \frac{c_p}{b_p^2 - b_p} \\ &= 1 + m - \frac{c_2}{b_2^2 - b_2}.\end{aligned}$$

This finishes the proof of (6.30) and shows that S (6.29) is a filtered λ-ring structure on $\mathbf{Z}[x]/\langle x^3 \rangle$.

Next we claim that S is isomorphic to R. Using Lemma 6.28 it remains to prove (6.25). If $r \leq b_2(b_2-1)/2$, then

$$c_2 - \overline{c}_2 = c_2 - r = mb_2(b_2 - 1).$$

On the other hand, if $r > b_2(b_2-1)/2$, then

$$c_2 + \overline{c}_2 = c_2 + b_2(b_2 - 1) - r = (1+m)b_2(b_2 - 1).$$

In either case (6.25) is true. Thus, S is isomorphic to R.

Finally, we claim that

$$S = S((b_p), k) \tag{6.31}$$

for some odd integer k satisfying (6.20). To prove (6.31) we use the notation

$$G_{2,p} = \gcd\left(b_2^2 - b_2, b_p^2 - b_p\right)$$

for any prime p. From the equation $(b_2^2 - b_2)c_p = (b_p^2 - b_p)c_2$, we obtain

$$\frac{b_2^2 - b_2}{G_{2,p}} c_p = \frac{b_p^2 - b_p}{G_{2,p}} c_2.$$

If $b_p \neq 0$, then this implies that $(b_2^2 - b_2)/G_{2,p}$ divides c_2. Since $(b_2^2 - b_2)/G_{2,p}$ clearly divides $(b_2^2 - b_2)$, it follows that $(b_2^2 - b_2)/G_{2,p}$ divides r. This in turn implies that $\bar{c}_2$ is divisible by $(b_2^2 - b_2)/G_{2,p}$, so we have

$$\bar{c}_2 G_{2,p} \equiv 0 \pmod{b_2^2 - b_2}. \tag{6.32}$$

The condition (6.32) clearly still holds when $b_p = 0$. Thus, we have $\bar{c}_2 G \equiv 0 \pmod{b_2^2 - b_2}$, where G is defined in (6.17). Equivalently, we have $\bar{c}_2 = k(b_2^2 - b_2)/G$ for some integer k. From the condition (6.30), we also have

$$\bar{c}_p = \frac{\bar{c}_2}{b_2^2 - b_2}(b_p^2 - b_p) = \frac{k(b_p^2 - b_p)}{G}.$$

It remains to show that k is odd and satisfies (6.20).

Using Lemma 6.27 (2), we know that $2^{\nu_2(b_2)}$ divides G and is, therefore, equal to $2^{\nu_2(G)}$. Since $(b_2 - 1)$ and $\bar{c}_2$ are both odd, it follows that k is odd as well. Also, k is positive because $\bar{c}_2$, $(b_2^2 - b_2)$, and G are all positive. Next observe that

$$1 \leq \bar{c}_2 \leq \frac{b_2^2 - b_2}{2}$$

from the definition (6.28). Multiplying by $G/(b_2^2 - b_2)$ we see that

$$1 \leq k \leq \frac{G}{2},$$

proving the first condition in (6.20).

Finally, let p be an odd prime satisfying (6.16). In this case, we have $\nu_p(G) = \nu_p(b_p)$, so

$$\nu_p(k) = \nu_p\left(\frac{k(b_p^2 - b_p)}{G}\right) = \nu_p(\bar{c}_p).$$

Since $\bar{c}_p$ is divisible by p, it follows that k is divisible by p as well. This proves the second condition in (6.20). This finishes the proof that R is isomorphic to $S((b_p), k) = S$. □

We obtain Theorems 6.19 and 6.21 by combining Lemmas 6.26, 6.29, 6.30, and 6.31.

In the next example, we show how the λ-operations on the filtered λ-rings $S((c_p))$ can be computed. Computing the λ-operations on $S((b_p), k)$ is left as an exercise for the reader.

Example 6.32. Consider the filtered λ-ring structure

$$S((c_p)) = \{\psi^p(x) = c_p x^2 : p \text{ prime}\}$$

on $\mathbf{Z}[x]/\langle x^3 \rangle$ as in (6.14). A general element in $S((c_p))$ has the form $a_0 + a_1 x + a_2 x^2$ for some integers a_0, a_1, and a_2. Applying λ_t we obtain

$$\begin{aligned}\lambda_t(a_0 + a_1 x + a_2 x^2) &= \lambda_t(a_0)\lambda_t(x)^{a_1}\lambda_t(x^2)^{a_2} \\ &= (1+t)^{a_0}\lambda_t(x)^{a_1}\lambda_t(x^2)^{a_2}.\end{aligned}$$

Thus, to describe the λ-operations on $S((c_p))$, it suffices to describe $\lambda_t(x)$ and $\lambda_t(x^2)$. We claim that

$$\begin{aligned}\lambda_t(x) &= 1 + xt + \left(\frac{x^2 - c_2x^2}{2}\right)t^2 + \sum_{p \text{ odd}} \left(\frac{c_px^2}{p}\right)t^p, \\ \lambda_t(x^2) &= 1 + x^2t,\end{aligned} \tag{6.33}$$

where the sum is taken over all the odd primes p.

To prove the first statement in (6.33), first observe that by (3.3) and $x^3 = 0$, we have $\psi^n = 0$ if $n \neq 1$ or a prime. This implies that

$$\begin{aligned}\psi_{-t}(x) &= \sum_{n\geq 1}(-1)^n\psi^n(x)t^n \\ &= -xt + \sum_{p \text{ prime}} (-1)^p(c_px^2)t^p.\end{aligned}$$

Dividing by $-t$ and integrating with respect to t, we obtain

$$\log \lambda_t(x) = xt + \sum_{p \text{ prime}} (-1)^{p-1}\frac{c_px^2}{p}t^p.$$

Exponentiating both sides and using $x^3 = 0$, we have

$$\begin{aligned}\lambda_t(x) &= \exp\left(xt + \sum_{p \text{ prime}} (-1)^{p-1}\frac{c_px^2}{p}t^p\right) \\ &= \exp(xt) \times \prod_{p \text{ prime}} \exp\left((-1)^{p-1}\frac{c_px^2}{p}t^p\right) \\ &= \left(1 + xt + \frac{x^2}{2}t^2\right) \times \prod_{p \text{ prime}} \left(1 + (-1)^{p-1}\frac{c_px^2}{p}t^p\right) \\ &= \left(1 + xt + \frac{x^2}{2}t^2\right) \times \left(1 + \sum_{p \text{ prime}} (-1)^{p-1}\frac{c_px^2}{p}t^p\right) \\ &= 1 + xt + \frac{x^2}{2}t^2 + \sum_{p \text{ prime}} (-1)^{p-1}\frac{c_px^2}{p}t^p.\end{aligned}$$

This proves the first statement in (6.33).

For the second assertion in (6.33), note that for any prime p, $\psi^p(x^2) = \psi^p(x)^2 = 0$ because $x^3 = 0$. Thus, we have $\psi^n(x^2) = 0$ for all $n \geq 2$. Using the Newton Formula (Theorem 3.10) and $x^3 = 0$, we have

$$0 = 0 - \lambda^1(x^2)\psi^1(x^2) = -2\lambda^2(x^2),$$

which implies that $\lambda^2(x^2)$ is equal to zero. Pick an $n > 2$. Suppose by induction that $\lambda^k(x^2)$ is equal to zero for $2 \leq k < n$. In this case, the Newton Formula says that

$$0 = \sum_{i=0}^{n-1}(-1)^i\lambda^i(x^2)\psi^{n-i}(x^2) = (-1)^{n+1}n\lambda^n(x^2).$$

This again implies that $\lambda^n(x^2)$ is equal to zero. This proves the second assertion in (6.33).

6.5 Exercises

(1) Let A be a $\mathbf{Q}$-algebra (e.g., $\mathbf{Q}$ and $S \otimes \mathbf{Q}$ for any $\mathbf{Z}$-torsion-free ring S). Let R be one of the following (truncated) polynomial or power series filtered rings with the x-adic filtration:

  (a) $A[x]/\langle x^n \rangle$ for $2 \leq n < \infty$.
  (b) $A[x]$.
  (c) $A[[x]]$.

  Prove that R admits uncountably many mutually non-isomorphic filtered λ-ring structures.

(2) For the reader who knows some algebraic topology: Let X be a topological space with the homotopy type of a CW complex. Given a CW structure on X, there is an associated filtered λ-ring structure on its K-theory $K(X)$ (Example 6.9). Prove that any two choices of CW structures on X give rise to isomorphic filtered λ-ring structures on $K(X)$.

(3) Prove (6.12) using the Newton Formula (Theorem 3.10).

(4) For the filtered λ-rings $S((b_p), k)$ as in (6.19), compute $\lambda_t(x)$ and $\lambda_t(x^2)$.

(5) Find the exact number of isomorphism classes of filtered λ-rings of the forms $S((p^3), k)$ and $S((p^4), k)$. Describe the possible values of k in each case.

Chapter 7

Filtered λ-Rings, II

In this chapter we first consider the set of filtered λ-ring structures over a power series ring $A[[x]]$. This set is, in general, non-trivial. For example, it is known [Yau (2004b)] that the set $\lambda(\mathbf{Z}[[x_1, \ldots, x_n]])$ of isomorphism classes of filtered λ-ring structures over the n-variable power series ring is uncountable for every $n \geq 1$. Likewise, the power series ring $\mathbf{Q}[[x]]$ admits uncountably many isomorphism classes of filtered λ-ring structures (Exercise 1, Chapter 6).

In §7.1 we consider the function

$$A \mapsto \{\text{filtered } \lambda\text{-ring structures on } A[[x]]\}$$

that sends a ring A with $\mathbf{Z} \subseteq A \subseteq \mathbf{Q}$ to the set of filtered λ-ring structures on the power series ring $A[[x]]$ with the x-adic filtration. It is shown that this function is *corepresentable* and that the corepresenting ring is a binomial ring (Definition 5.1). In other words, there exists a binomial ring Γ with the following property. For every ring A with $\mathbf{Z} \subseteq A \subseteq \mathbf{Q}$, there is a canonical bijection

$$\zeta_A \colon \mathbf{Ring}(\Gamma, A) \xrightarrow{\cong} \{\text{filtered } \lambda\text{-ring structures on } A[[x]]\}$$

of sets. Here $\mathbf{Ring}(\Gamma, A)$ denotes the set of ring homomorphisms from Γ to A. This corepresenting ring Γ is highly non-trivial, since both $\mathbf{Z}[[x]]$ and $\mathbf{Q}[[x]]$ have uncountably many isomorphism classes of filtered λ-ring structures.

Note that this corepresentability result is very much in the spirit of formal group theory. A basic result about (one-dimensional and commutative) formal group laws is that the function

$$R \mapsto \{\text{formal group laws over } R\}$$

sending a ring R to the set of formal group laws over R is corepresentable. The corepresenting ring L is called the *Lazard ring*. The reader is referred to [Hazewinkel (1978)] for detailed discussion of formal group laws.

In §7.2 the universal λ-ring functor Λ (Chapter 2) is extended to the setting of filtered rings. Filtered λ-rings are then shown to be exactly the coalgebras of the resulting comonad on the category of filtered rings.

The reference for this Chapter is [Yau (2004a)].

7.1 Power Series Rings

The main result of this section is:

Theorem 7.1. *There exists a binomial ring* Γ *such that, for every subring* A *of* $\mathbf{Q}$, *there is a canonical bijection*

$$\zeta_A \colon \mathbf{Ring}(\Gamma, A) \xrightarrow{\cong} \{\text{filtered } \lambda\text{-ring structures on } A[[x]]\}$$

of sets.

In this Theorem, the power series ring $A[[x]]$ is given the x-adic filtration (Example 6.3), and elements in A have filtration 0. One way to rephrase Theorem 7.1 is that the function

$$A \mapsto \{\text{filtered } \lambda\text{-ring structures on } A[[x]]\}$$

sending any subring A of $\mathbf{Q}$ to the set of filtered λ-ring structures on $A[[x]]$ is *corepresentable.* The corepresenting object is a certain universal ring Γ (7.7), which turns out to be a binomial ring (Proposition 7.4). The bijection ζ_A is defined in (7.12).

Recall from Corollary 6.12 that a filtered λ-ring structure on $A[[x]]$ is uniquely determined by the power series $\psi^p(x) \in A[[x]]$, where p runs through the set of primes. These power series have to satisfy the following conditions:

(1) The constant term in each $\psi^p(x)$ is 0.
(2) $\psi^p(\psi^q(x)) = \psi^q(\psi^p(x))$ for any primes p and q.
(3) $\psi^p(x) \equiv x^p \pmod{pA[[x]]}$.

Thus, we need the universal examples of the commuting power series $\psi^p_\Gamma(x)$ (7.9) over Γ with 0 constant term that satisfy the mod p condition. The bijection ζ_A is then obtained by sending a ring homomorphism $f \colon \Gamma \to A$ to the power series $f(\psi^p_\Gamma(x)) \in A[[x]]$, which constitute a filtered λ-ring structure on $A[[x]]$.

We need the following preliminary result for the construction of Γ.

Definition 7.2. In a ring R, an element $r \in R$ is called a $\mathbf{Z}$-*torsion* if $nr = 0$ for some non-zero integer n.

Of course, a ring is $\mathbf{Z}$-torsion-free (Definition 3.12) if and only if the only $\mathbf{Z}$-torsion element in R is 0. Since 0 is always a $\mathbf{Z}$-torsion element, the set of $\mathbf{Z}$-torsion elements in a ring R is non-empty.

Lemma 7.3. *Let* R *be a ring, and let* J *be the set of* $\mathbf{Z}$-*torsion elements in* R. *Then* J *is an ideal in* R. *Moreover, the quotient ring* R/J *is* $\mathbf{Z}$-*torsion-free.*

Proof. It is clear that $0 \in R$ is a $\mathbf{Z}$-torsion element. Pick two elements a and b in J. Then $na = 0 = mb$ for some non-zero integers n and m. It follows that $nm(a + b) = (na)m + n(mb) = 0$, which implies that $a + b$ is an element of J.

Finally, pick an element $r \in R$. Then we have $n(ar) = (na)r = 0$, which implies that $ar \in J$. This shows that the set J of **Z**-torsion elements in a ring R is an ideal.

To see that the quotient ring R/J is **Z**-torsion-free, suppose that $r + J$ is an element in R/J and

$$n(r + J) = 0 + J \tag{7.1}$$

for some non-zero integer n. We want to show that $r + J = J$, or equivalently, $r \in J$. From (7.1) it follows that $nr \in J$, so there exists a non-zero integer m such that $mnr = 0$. Since $mn \neq 0$ we conclude that $r \in J$. □

7.1.1 *The universal ring* Γ

Now we begin the construction of the universal ring Γ.

For integers $n \geq 0$, $i \geq 1$, and primes $p, q_1, \ldots, q_n$, let $x_{(p,i,q_1,\ldots,q_n)}$ be independent variables. Let X be the set of such variables. Consider the polynomial ring $\mathbf{Z}[X] = \mathbf{Z}\left[\{x_{(p,i,q_1,\ldots,q_n)}\}\right]$. We define some elements $u_{(p,i)}$, $w_{(p,q,k)}$, and $V(p,i,q_1,\ldots,q_{n+1})$ in this polynomial ring as follows.

For an integer $i \geq 1$ and a prime p, we define

$$u_{(p,i)} = \begin{cases} px_{(p,i)} & \text{if } i \neq p, \\ 1 + px_{(p,p)} & \text{if } i = p. \end{cases} \tag{7.2}$$

Let t be a variable. For an integer $k \geq 1$ and primes p and q, we define $w_{(p,q,k)}$ by the equality

$$\sum_{i\geq 1} u_{(p,i)} \Big(\sum_{j\geq 1} u_{(q,j)} t^j\Big)^i - \sum_{i\geq 1} u_{(q,i)} \Big(\sum_{j\geq 1} u_{(p,j)} t^j\Big)^i = \sum_{k\geq 1} w_{(p,q,k)} t^k. \tag{7.3}$$

For integers $i \geq 1$, $n \geq 0$, and primes $p, q_1, \ldots, q_{n+1}$, we define

$$\begin{aligned} V(p,i,q_1,\ldots,q_{n+1}) = x^{q_{n+1}}_{(p,i,q_1,\ldots,q_n)} &- x_{(p,i,q_1,\ldots,q_n)} \\ &- q_{n+1}x_{(p,i,q_1,\ldots,q_{n+1})}. \end{aligned} \tag{7.4}$$

In the polynomial ring $\mathbf{Z}[X]$, let I be the ideal

$$I = \left\langle w_{(p,q,k)}, V(p,i,q_1,\ldots q_{n+1})\right\rangle \tag{7.5}$$

generated by the elements $w_{(p,q,k)}$ and $V(p,i,q_1,\ldots q_{n+1})$ for all possible integers $n \geq 0$, $i, k \geq 1$, and primes $p, q, q_1, \ldots q_{n+1}$.

Let J be the ideal of **Z**-torsion elements in the quotient ring

$$R = \mathbf{Z}[X]/I. \tag{7.6}$$

Define Γ as the quotient ring

$$\Gamma = R/J = \frac{\mathbf{Z}[X]/I}{\langle \mathbf{Z}\text{-torsion}\rangle}. \tag{7.7}$$

In what follows, we will write $x_{(p,i,q_1,\ldots,q_n)}$ and $u_{(p,i)}$ for their images in R and Γ.

Proposition 7.4. *The ring* Γ (7.7) *is a binomial ring.*

Proof. By Lemma 7.3 we know that Γ is $\mathbf{Z}$-torsion-free. Thus, by Corollary 5.11 it remains to show that

$$a^p \equiv a \pmod{p\Gamma} \tag{7.8}$$

for elements $a \in \Gamma$ and prime p. It suffices to prove (7.8) for the ring R (7.6) instead of Γ, since Γ is a quotient of R. Since R is generated as a ring by the variables $x_{(p,i,q_1,\ldots,q_n)}$, it suffices to prove

$$x^{q_{n+1}}_{(p,i,q_1,\ldots,q_n)} \equiv x_{(p,i,q_1,\ldots,q_n)} \pmod{q_{n+1}R}$$

for an arbitrary prime q_{n+1}. From (7.4) we see that

$$x^{q_{n+1}}_{(p,i,q_1,\ldots,q_n)} - x_{(p,i,q_1,\ldots,q_n)} = q_{n+1}x_{(p,i,q_1,\ldots,q_{n+1})}$$

for every variable $x_{(p,i,q_1,\ldots,q_n)}$ in R and every prime q_{n+1}. This is exactly what we wanted. □

7.1.2 *The universal* ψ^p

Let x be a variable. We now define the universal power series $\psi^p_\Gamma(x)$ over Γ. For each prime p, we define

$$\psi^p_\Gamma(x) = \sum_{i=1}^{\infty} u_{(p,i)}x^i = \left(1 + px_{(p,p)}\right)x^p + \sum_{i\geq 1,\, i\neq p} px_{(p,i)}x^i \in \Gamma[[x]]. \tag{7.9}$$

Lemma 7.5. *The following statements hold for the power series* $\psi^p_\Gamma(x)$ (7.9) *in* $\Gamma[[x]]$ *for any primes* p *and* q.

(1) $\psi^p_\Gamma(x)$ *has constant term* 0.
(2) $\psi^p_\Gamma(\psi^q_\Gamma(x)) = \psi^q_\Gamma(\psi^p_\Gamma(x))$.
(3) $\psi^p_\Gamma(x) \equiv x^p \pmod{p\Gamma[[x]]}$.

Proof. Statements (1) and (3) are immediate from the definition (7.9). For statement (2), note that

$$\psi^p_\Gamma(\psi^q_\Gamma(x)) = \sum_{i\geq 1} u_{(p,i)}\left(\sum_{j\geq 1} u_{(q,j)}x^j\right)^i$$

Thus, the required equality $\psi^p_\Gamma(\psi^q_\Gamma(x)) = \psi^q_\Gamma(\psi^p_\Gamma(x))$ in $\Gamma[[x]]$ follows from (7.3) and the definition (7.7) of Γ. □

Let A be a subring of the rationals $\mathbf{Q}$, and let $f\colon \Gamma \to A$ be a ring homomorphism. For each prime p, consider the power series

$$\begin{aligned} f(\psi^p_\Gamma(x)) &= \sum_{i=1}^{\infty} f\left(u_{(p,i)}\right)x^i \\ &= \left(1 + pf\left(x_{(p,p)}\right)\right)x^p + \sum_{i\geq 1,\, i\neq p} pf\left(x_{(p,i)}\right)x^i \in A[[x]]. \end{aligned} \tag{7.10}$$

In other words, $f(\psi^p_\Gamma(x))$ is obtained from $\psi^p_\Gamma(x)$ by applying f to the coefficients of x^i $(i \geq 1)$.

Lemma 7.6. *The power series $f(\psi^p_\Gamma(x)) \in A[[x]]$ are the Adams operations (when applied to x) of a unique filtered λ-ring structure on $A[[x]]$.*

Proof. From the definition (7.10), the three statements in Lemma 7.5 are also satisfied for $f(\psi^p_\Gamma(x))$. Thus, the Lemma follows from Corollary 6.12. □

The following result shows that any ring homomorphism from Γ to a subring of **Q** is uniquely determined by its action on the $x_{(p,i)}$.

Lemma 7.7. *Let $f\colon \Gamma \to A$ be a ring homomorphism for some subring $A \subseteq \mathbf{Q}$. Then*

$$f\left(x_{(p,i,q_1,\ldots,q_{n+1})}\right) = \frac{f\left(x_{(p,i,q_1,\ldots,q_n)}\right)^{q_{n+1}} - f\left(x_{(p,i,q_1,\ldots,q_n)}\right)}{q_{n+1}} \tag{7.11}$$

for all integers $n \geq 0$, $i \geq 1$, and primes $p, q_1, \ldots, q_{n+1}$. In particular, f is completely determined by $f\left(x_{(p,i)}\right)$ for $i \geq 1$ and p prime.

Proof. We know that $V_{(p,i,q_1,\ldots,q_{n+1})} = 0$ in Γ by construction (7.7). From (7.4) we see that

$$q_{n+1} f\left(x_{(p,i,q_1,\ldots,q_{n+1})}\right) = f\left(x_{(p,i,q_1,\ldots,q_n)}\right)^{q_{n+1}} - f\left(x_{(p,i,q_1,\ldots,q_n)}\right).$$

The equality (7.11) now follows because A is **Z**-torsion-free. □

7.1.3 *The bijection ζ*

Let A be a subring of **Q**. Now we define the function

$$\zeta_A\colon \mathbf{Ring}(\Gamma, A) \to \{\text{filtered } \lambda\text{-ring structures on } A[[x]]\} \tag{7.12}$$

by setting

$$\zeta_A(f) = \{f(\psi^p_\Gamma(x))\}$$

for $f \in \mathbf{Ring}(\Gamma, A)$. Here $\{f(\psi^p_\Gamma(x))\}$ is the unique filtered λ-ring structure on $A[[x]]$ determined by the power series $f(\psi^p_\Gamma(x))$ (Lemma 7.6).

We can now finish the proof of Theorem 7.1.

Proof. [Proof of Theorem 7.1] It is shown in Proposition 7.4 that Γ is a binomial ring. It remains to show that ζ_A (7.12) is a bijection. To see that ζ_A is injective, suppose that $\zeta_A(f) = \zeta_A(g)$ for some ring homomorphisms $f, g \in \mathbf{Ring}(\Gamma, A)$. This implies that $f\left(u_{(p,i)}\right) = g\left(u_{(p,i)}\right)$ for integers $i \geq 1$ and primes p. Since A is **Z**-torsion-free, it follows from the definition (7.2) of $u_{(p,i)}$ that $f\left(x_{(p,i)}\right) = g\left(x_{(p,i)}\right)$. Thus, by Lemma 7.7 f is equal to g, showing that ζ_A is injective.

To show that ζ_A is surjective, suppose that $A[[x]]$ has a given filtered λ-ring structure with Adams operations ψ^p. The power series $\psi^p(x) = \sum_{i=1}^\infty a_{(p,i)} x^i \in$

$A[[x]]$ (p primes) satisfy the three conditions in Corollary 6.12. In particular, we have

$$a_{(p,i)} \equiv \begin{cases} 0 \pmod{pA} & \text{if } i \neq p, \\ 1 \pmod{pA} & \text{if } i = p \end{cases} \tag{7.13}$$

We want to find a ring homomorphism $f \in \mathbf{Ring}(\Gamma, A)$ such that

$$f(\psi^p_\Gamma(x)) = \psi^p(x) = \sum_{i=1}^{\infty} a_{(p,i)} x^i. \tag{7.14}$$

We first construct a ring homomorphism from $\mathbf{Z}[X]$ to A.

On the generators $x_{(p,i)} \in \mathbf{Z}[X]$ for $i \geq 1$ and p prime, define

$$f\left(x_{(p,i)}\right) = \begin{cases} \dfrac{a_{(p,i)}}{p} & \text{if } i \neq p, \\ \dfrac{a_{(p,p)} - 1}{p} & \text{if } i = p. \end{cases} \tag{7.15}$$

This makes sense by (7.13) and the fact that A is $\mathbf{Z}$-torsion-free. The map f is extended to the other generators in $\mathbf{Z}[X]$ inductively by the formula:

$$f\left(x_{(p,i,q_1,\ldots,q_{n+1})}\right) = \frac{f\left(x_{(p,i,q_1,\ldots,q_n)}\right)^{q_{n+1}} - f\left(x_{(p,i,q_1,\ldots,q_n)}\right)}{q_{n+1}} \tag{7.16}$$

We now have a ring homomorphism $f \colon \mathbf{Z}[X] \to A$, which we claim descends to Γ.

From the definitions (7.15) of $f\left(x_{(p,i)}\right)$ and (7.2) of $u_{(p,i)}$, it follows that

$$f\left(u_{(p,i)}\right) = a_{(p,i)}. \tag{7.17}$$

Using the definition (7.3) of $w_{(p,q,k)}$, this implies that

$$\sum_{k\geq 1} f\left(w_{(p,q,k)}\right) x^k = \psi^p(\psi^q(x)) - \psi^q(\psi^p(x)) = 0.$$

Thus, f sends $w_{(p,q,k)}$ to $0 \in A$.

Likewise, we have

$$\begin{aligned} f\left(V(p, i, q_1, \ldots, q_{n+1})\right) &= f\left(x_{(p,i,q_1,\ldots,q_n)}\right)^{q_{n+1}} - f\left(x_{(p,i,q_1,\ldots,q_n)}\right) \\ &\quad - q_{n+1} f\left(x_{(p,i,q_1,\ldots,q_{n+1})}\right) \\ &= 0 \end{aligned}$$

by (7.16). Thus, f is trivial on the ideal I (7.5), which implies that f induces a ring homomorphism $f \colon R = \mathbf{Z}[X]/I \to A$. This f is trivial on the ideal J of $\mathbf{Z}$-torsion elements in R because A is $\mathbf{Z}$-torsion-free. So there is an induced ring homomorphism $f \colon \Gamma = R/J \to A$. This last f satisfies (7.14) because of (7.17). This finishes the proof that ζ_A is surjective.

The proof of Theorem 7.1 is complete. □

7.2 Filtered λ-Rings as Coalgebras

The purpose of this section is to extend the universal λ-ring comonad $(\Lambda, \lambda, \varepsilon)$ (Theorem 2.25) to the category of filtered rings (Theorem 7.10). The resulting comonad Λ^f has as its coalgebras the filtered λ-rings (Corollary 7.11).

First we extend the universal λ-ring functor Λ to the category $\mathbf{Ring}^f$ of filtered rings. For a ring R, recall from Chapter 2 that the underlying set of the universal λ-ring is

$$\Lambda(R) = \left\{1 + \sum_{i\geq 1} a_i t^i : a_i \in R\right\},$$

the set of strict power series over R. Its addition $+_\Lambda$ and multiplication $\times_\Lambda$ are defined in (2.4), and Theorem 2.5 shows that $\Lambda(R)$ is a ring. The λ-operations in $\Lambda(R)$ are defined in (2.8).

Let R be a filtered ring with filtration ideals I_n $(n \geq 0)$ (Definition 6.1). Define the subset

$$\Lambda(I^n) = \left\{1 + \sum_{i\geq 1} a_i t^i \in \Lambda(R) \colon a_i \in I^n \text{ for each } i \geq 1\right\} \subseteq \Lambda(R). \tag{7.18}$$

We now observe that these $\Lambda(I_n)$ make $\Lambda(R)$ into a filtered ring.

Lemma 7.8. *Let R be a filtered ring with filtration ideals I_n $(n \geq 0)$. Then the ring $\Lambda(R)$ is a filtered ring when equipped with the filtration ideals $\Lambda(I_n)$ (7.18).*

Proof. Since I_n contains I_{n+1}, it follows that $\Lambda(I_n)$ contains $\Lambda(I_{n+1})$. Also, $I_0 = R$ implies that $\Lambda(I_0) = \Lambda(R)$. It remains to show that each $\Lambda(I_n)$ is an ideal in the ring $\Lambda(R)$.

The additive zero in $\Lambda(R)$ is the power series 1, which lies in every $\Lambda(I_n)$ because $0 \in I_n$. To see that $\Lambda(I_n)$ is closed under addition, pick two elements $e = 1 + \sum_{i\geq 1} a_i t^i$ and $g = 1 + \sum_{i\geq 1} b_i t^i$ in $\Lambda(I_n)$. Then their sum in $\Lambda(R)$ is

$$e +_\Lambda g = 1 + \sum_{n\geq 1} \left(\sum_{i+j=n} a_i b_j\right) t^n,$$

where we set $a_0 = b_0 = 1$. Since $a_i, b_i \in I_n$ for $i \geq 1$, it follows that $e +_\Lambda g$ lies in $\Lambda(I_n)$.

Finally, pick an element $h = 1 + \sum_{i\geq 1} c_i t^i \in \Lambda(R)$. We must show that the product $e \times_\Lambda h$ lies in I_n. For $i \geq 1$ the coefficient of t^i in $e \times_\Lambda h$ is $P_i(a_1, \ldots, a_i; c_1, \ldots, c_i)$. Every summand in $P_i(a_1, \ldots, a_i; c_1, \ldots, c_i)$ must contain a factor of a_j for some $j \in \{1, \ldots, i\}$ (Chapter 1 Exercise 6). Since I_n is an ideal in R, this implies that $P_i(a_1, \ldots, a_i; c_1, \ldots, c_i) \in I_n$. Thus, the product $e \times_\Lambda h$ lies in I_n. □

Proposition 7.9. *There is a functor $\Lambda^f \colon$ $\mathbf{Ring}^f \to \mathbf{Ring}^f$ on the category $\mathbf{Ring}^f$ of filtered rings such that:*

(1) *For a filtered ring R with filtration ideals I_n, $\Lambda^f(R)$ is the filtered ring $\Lambda(R)$ with filtration ideals $\Lambda(I_n)$.*
(2) *For a filtered ring homomorphism φ, $\Lambda^f(\varphi)$ coincides with $\Lambda(\varphi)$ on the underlying rings.*

Proof. From Lemma 7.8 we know that $\Lambda(R)$ is a filtered ring with filtration ideals $\Lambda(I_n)$. Since Λ itself is a functor on the category of rings (Theorem 2.25), it remains to show that $\Lambda^f(\varphi)\colon \Lambda^f(R) \to \Lambda^f(S)$ is a filtered ring homomorphism if $\varphi\colon R \to S$ is a filtered ring homomorphism. By definition we have

$$\Lambda^f(\varphi)(1 + a_1t + a_2t^2 + \cdots) = 1 + \varphi(a_1)t + \varphi(a_2)t^2 + \cdots .$$

If every a_i lies in I_n^R for some fixed n, then every $\varphi(a_i)$ lies in I_n^S, so $\Lambda^f(\varphi)$ is a filtered ring homomorphism. □

Recall from Theorem 2.25 that the universal λ-ring comonad $(\Lambda, \lambda, \varepsilon)$ has comultiplication $\lambda\colon \Lambda \to \Lambda^2$ and counit $\varepsilon\colon \Lambda \to Id$. On an element $g = 1 + \sum_{i\geq 1} a_it^i \in \Lambda(R)$, they are given by

$$\lambda_R(g) = 1 + \sum_{m\geq 1}\left(1 + \sum_{n\geq 1} P_{n,m}(a_1, \ldots, a_{nm})t^n\right)t^m \in \Lambda^2(R),$$

$$\varepsilon_R(g) = a_1 \in R.$$

Theorem 7.10. *There is a comonad $(\Lambda^f, \lambda^f, \varepsilon^f)$ on the category* $\mathbf{Ring}^f$ *of filtered rings such that, for a filtered ring R,*

$$\lambda_R^f\colon \Lambda^f(R) \to \Lambda^f(\Lambda^f(R)) \quad \textit{and} \quad \varepsilon_R^f\colon \Lambda^f(R) \to R$$

coincide with λ_R and ε_R, respectively, on the underlying rings.

Proof. We already know that Λ^f is a functor on $\mathbf{Ring}^f$ (Proposition 7.9) and that $(\Lambda, \lambda, \varepsilon)$ is a comonad on **Ring** (Theorem 2.25). Thus, it remains to show that λ_R^f and ε_R^f are filtered ring homomorphisms for every filtered ring R. This is clear for ε_R^f.

For λ_R^f, suppose that $g \in \Lambda(I_k)$, where I_k is the kth filtration ideal in R. Then the coefficient a_i of t^i $(i \geq 1)$ in g lies in I_k. Since the polynomial $P_{n,m}$ has no constant terms, it follows that $P_{n,m}(a_1, \ldots, a_{nm})$ lies in I_k for $n, m \geq 1$. This shows that $\lambda_R^f(g)$ lies in the kth filtration ideal in $\Lambda^f(\Lambda^f(R))$. □

Now we use the comonad $(\Lambda^f, \lambda^f, \varepsilon^f)$ to describe filtered λ-rings.

Corollary 7.11. *The coalgebras of the comonad $(\Lambda^f, \lambda^f, \varepsilon^f)$ on the category* $\mathbf{Ring}^f$ *of filtered rings are exactly the filtered λ-rings.*

Proof. Let R be a filtered λ-ring. By Theorem 2.26 the λ-ring structure on R (without the filtration) corresponds to a Λ-coalgebra structure (Definition 2.21) with structure map $\lambda_t\colon R \to \Lambda(R)$. To see that the filtered ring R is a Λ^f-coalgebra, it suffices to show that λ_t is a filtered ring homomorphism. So pick an element

$r \in I_k$, the kth filtration ideal in R. Then $\lambda^i(r) \in I_k$ for every $i \geq 1$, so $\lambda_t(r) = 1 + \sum_{i\geq 1} \lambda^i(r)t^i \in \Lambda(I_k)$. This shows that λ_t preserves the filtration ideals.

Conversely, suppose that R is a Λ^f-coalgebra with structure map $\rho\colon R \to \Lambda^f(R)$. By Theorem 7.10 the ring R (without the filtration) is a Λ-coalgebra with structure map ρ. Thus, ρ corresponds to a λ-ring structure on R (Theorem 2.26) with $\rho(r) = 1 + \sum_{i\geq 1} \lambda^i(r)t^i$. To see that the filtration ideal I_k in R is a λ-ideal, pick an $r \in I_k$. Since ρ is a filtered ring homomorphism, we have $\rho(r) \in \Lambda(I_k)$, which implies that $\lambda^i(r) \in I_k$ for all $i \geq 1$. So I_k is a λ-ideal. □

Here is another consequence of Theorem 7.10.

Corollary 7.12. *Let R be a filtered λ-ring with filtration ideals I_n. Then $\Lambda^f(R)$ is a filtered λ-ring.*

Proof. We already know that $\Lambda^f(R)$ is a filtered ring with filtration ideals $\Lambda(I_n)$ (Lemma 7.8) and that it is a λ-ring (Theorem 2.6). In the proof of Theorem 7.10 it is shown that the λ-operations in $\Lambda^f(R)$ preserves the filtration ideals $\Lambda(I_n)$. □

7.3 Exercises

(1) Let $(R, \{I_n\})$ be a filtered ring. It is said to be *complete* (respectively, *Hausdorff*) if and only if the natural ring homomorphism $p\colon R \to \varprojlim R/I_n$ induced by the projections $R \to R/I_n$ is surjective (respectively, injective). Prove that $(R, \{I_n\})$ is complete (respectively, Hausdorff) if and only if $(\Lambda(R), \{\Lambda(I_n)\})$ is complete (respectively, Hausdorff).

(2) Recall the big Witt vectors functor $\mathbf{W}$ from Corollary 4.12. For a filtered ring $(R, \{I_n\})$, define the subset

$$\mathbf{W}(I_n) = \{(a_1, a_2, \ldots) \in \mathbf{W}(R)\colon \text{each } a_i \in I_n\} \subseteq \mathbf{W}(R)$$

for each $n \geq 1$. Prove the following statements.

(a) $\mathbf{W}(R)$ is a filtered ring with filtration ideals $\mathbf{W}(I_n)$.
(b) $\mathbf{W}$ extends to a functor $\mathbf{W}^f$ on the category $\mathbf{Ring}^f$ of filtered rings.
(c) The Witt polynomial w_n extends to a natural transformation

$$w_n\colon \mathbf{W}^f \to Id$$

of functors on the category of filtered rings.
(d) Extend Corollary 4.13 to the setting of filtered rings.
(e) The Artin-Hasse Exponential isomorphism (Theorem 4.16) extends to a natural isomorphism $E^f\colon \Lambda^f \to \mathbf{W}^f$ of functors on the category of filtered rings.

(3) Prove that the natural transformation z_n (Theorem 4.21) extends to a natural transformation $z_n\colon \Lambda^f \to Id$ of functors on the category of filtered rings. Then extend Corollary 4.22 to the setting of filtered rings.

(4) Let $f(x)$ and $g(x)$ be two power series over a field $\mathbf{k}$ such that $f(x) \equiv ax \equiv g(x) \pmod{x^2}$ for some element $a \in \mathbf{k}$ that is neither 0 nor a root of unity. Prove: For every element $c \in \mathbf{k}$, there exists a unique power series $h(x)$ over $\mathbf{k}$ satisfying the following two conditions:

(a) $h(x) \equiv cx \pmod{x^2}$.

(b) $h(g(x)) = f(h(x))$ as power series.

(5) Let R be a subring of $\mathbf{Q}$, and let R_1 and R_2 be two filtered λ-ring structures on the power series ring $R[[x]]$ with the x-adic filtration. Suppose that for each prime p, there exists an element $r_p \in R$, which is neither 0 nor a root of unity, such that $\psi_1^p(x) \equiv r_p x \equiv \psi_2^p(x) \pmod{x^2}$, where ψ_1^p and ψ_2^p are the Adams operations in R_1 and R_2, respectively. Let $\varphi \colon R_1 \to R_2$ be a filtered ring homomorphism. Prove: φ is a filtered λ-homomorphism if and only if there exists a prime q such that $\varphi\psi_1^q = \psi_2^q\varphi$. This is called the *Hasse Principle* in [Yau (2004a)].

Chapter 8

Open Problems

In this final chapter we provide a list of open problems related to λ-rings. Purely algebraic questions are listed first, followed by problems about representation rings and topological K-theory.

8.1 Power Series and Polynomial Rings

It was mentioned in Chapter 6 that the power series ring $\mathbf{Z}[[x_1, \ldots, x_n]]$ on n variables ($n \geq 1$) admits uncountably many isomorphism classes of filtered λ-ring structures. This result is proved in [Yau (2004b)] using a combination of results about the topological K-theory of spaces in the Mislin genus of the classifying space of a connected compact Lie group. In particular, its only known proof is mainly topological/homotopical, although the statement itself is purely algebraic. This leads us to the first problem.

Problem 8.1. *Show by purely algebraic methods that the power series ring $\mathbf{Z}[[x_1, \ldots, x_n]]$ on n variables ($n \geq 1$) admits uncountably many isomorphism classes of filtered λ-ring structures.*

The case when at least one of the x_i is nilpotent ($x_i^{e_i} = 0$ for some $e_i < \infty$) is Theorem 6.13. Perhaps there is a proof similar to that of Theorem 6.13 that works for the n-variable power series rings.

The next problem is about classification of λ-ring structures on a given ring.

Problem 8.2. *Classify the isomorphism classes of filtered λ-ring structures on the filtered power series or polynomial rings*

$$R = \frac{A[[x_1, \ldots, x_n]]}{\langle x_1^{e_1}, \ldots, x_n^{e_n} \rangle} \quad or \quad \frac{A[x_1, \ldots, x_n]}{\langle x_1^{e_1}, \ldots, x_n^{e_n} \rangle}$$

with the adic filtration (Example 6.4), for any subring $A \subseteq \mathbf{Q}$, $n \geq 1$, and exponents e_i with $2 \leq e_i \leq \infty$.

From Theorem 6.13 it is known that there are uncountably many such isomorphism classes when at least one e_j is finite. If all the exponents e_i are infinite (i.e., R is the power series ring on n variables) and $A = \mathbf{Z}$, then there are again uncountably many such isomorphism classes [Yau (2004b)]. Note that the case of the polynomial ring $\mathbf{Z}[x]$ was also studied in [Clauwens (1994)].

The two smallest cases of this classification problem, $R = \mathbf{Z}[x]/\langle x^2 \rangle$ or $\mathbf{Z}[x]/\langle x^3 \rangle$, are done in Corollary 6.17 and Theorems 6.19 and 6.21, respectively. For starters, one can consider the truncated polynomial rings $R = \mathbf{Z}[x]/\langle x^n \rangle$ for $3 < n < \infty$.

There is some recent interest in λ-rings from the perspective of class field theory [Borger and de Smit (2008)]. In this context, the truncated polynomial ring $\mathbf{Z}[\mu_r] = \mathbf{Z}[x]/\langle x^r - 1 \rangle$, where r is some positive integer, seems particularly important.

Problem 8.3. *Classify the isomorphism classes of λ-ring structures on the truncated polynomial ring $\mathbf{Z}[\mu_r] = \mathbf{Z}[x]/\langle x^r - 1 \rangle$, where $r \geq 2$ is a positive integer.*

There is at least one λ-ring structure on this truncated polynomial ring. Namely, by Wilkerson's Theorem 3.54 there is a λ-ring structure on $\mathbf{Z}[\mu_r]$ in which the Adams operations are given by $\psi^n(x) = x^n$ for $n \geq 1$ [Borger and de Smit (2008)].

8.2 Universal Ring Γ

Recall from Theorem 7.1 that there is a binomial ring Γ (Proposition 7.4) with the following property. For any subring A of the rationals $\mathbf{Q}$, the set of filtered λ-ring structures on the power series ring $A[[x]]$ is classified by the set $\mathbf{Ring}(\Gamma, A)$ of ring homomorphisms from Γ to A. Besides the construction (7.7) and the fact that it is binomial, we do not know much about the ring Γ. To better under the filtered λ-ring structures on the power series ring $A[[x]]$, it would be nice to know more about the universal ring Γ.

Problem 8.4. *Describe the structure of the universal ring* Γ.

Since Γ is a binomial ring (Proposition 7.4), it is the homomorphic image of a ring $\mathrm{Int}(\mathbf{Z}^X)$ of integer-valued polynomials for some set X (Theorem 5.34). One way to approach Problem 8.4 is to ask the question: Is Γ isomorphic to a ring of integer-valued polynomials for some set of variables? As a start, one can try to find a minimal set Y of variables such that Γ is the homomorphic image of the ring $\mathrm{Int}(\mathbf{Z}^Y)$ of integer-valued polynomials over Y.

The universal ring Γ is constructed such that there is a canonical bijection

$$\zeta_A \colon \mathbf{Ring}(\Gamma, A) \xrightarrow{\cong} \{\text{filtered } \lambda\text{-ring structures on } A[[x]]\}$$

for subrings A of $\mathbf{Q}$. It would be nice if the restriction $A \subseteq \mathbf{Q}$ can be removed.

Problem 8.5. *Construct a generalization of the universal ring Γ that works for all rings of characteristic* 0, *or show that it does not exist.*

Another way to try to generalize Γ is to consider more variables.

Problem 8.6. *Construct a ring $\overline{\Gamma}$ such that there is a canonical isomorphism*

$$\zeta_A \colon \mathbf{Ring}(\overline{\Gamma}, A) \xrightarrow{\cong} \{\text{filtered } \lambda\text{-ring structures on } A[[x_1, \ldots, x_n]]\}$$

for any subring A of $\mathbf{Q}$, where $x_1, \ldots, x_n$ are independent variables. Describe the structure of $\overline{\Gamma}$.

8.3 Representation Ring

Recall from Example 1.32 that the representation ring $R(G)$ of a group G is an augmented λ-ring. The augmentation sends an actual representation to its dimension.

A major result of [Osse (1997)] gives an intrinsic characterization of an augmented λ-ring that makes it isomorphic to the representation ring of a connected compact Lie group. Osse called an augmented λ-ring R *normal* if it satisfies the following conditions:

(1) R is finite-dimensional as a λ-ring.
(2) Every element of dimension one in R is invertible.
(3) R is finitely generated as a ring.
(4) R is an integrally closed domain.

It is shown in Theorem 3.1 in [Osse (1997)] that an augmented λ-ring R is isomorphic to the representation ring $R(G)$ of a connected compact Lie group G if and only if:

(1) R is normal, and
(2) the associated graded ring $Gr_\gamma(R) \otimes \mathbf{Q}$ is a polynomial ring over $\mathbf{Q}$.

Here

$$Gr_\gamma(R) = \bigoplus_{n \geq 0} I_n/I_{n+1} = (R/I_1) \oplus (I_1/I_2) \oplus \cdots$$

denotes the associated graded ring of R with respect to its γ-filtration $\{I_n\}$.

The next problem is the finite group version of Osse's Theorem.

Problem 8.7. *Give an intrinsic characterization of an augmented λ-ring that makes it isomorphic to the representation ring of a finite group.*

In this Problem, it may be more interesting to replace the class of finite groups by other classes of groups. For example, one can consider the class of hyperbolic groups [Gromov (1987)].

In the paper [Osse (1997)], Osse also showed that a connected compact Lie group G is determined by the λ-ring $R(G)$ together with the subgroup $RO(G)$ (or $RSp(G)$). Here $RO(G)$ and $RSp(G)$ are the *real* and *symplectic* representation rings of G.

Problem 8.8. *For a finite group G, to what extend does the λ-ring $R(G)$ determine the group G? In particular, is G determined by the λ-ring $R(G)$ together with some subgroups?*

Again, it may be more interesting in this Problem to replace the class of finite groups by other classes of groups.

8.4 Topological K-Theory

A λ-ring R is said to be *topologically realizable* if there exists a topological space X whose K-theory λ-ring $K(X)$ is isomorphic to R. For example, for any integer $n \geq 1$, the K-theory ring of the classifying space $BSp(n)$ of the symplectic group $Sp(n)$ is the power series ring $K(BSp(n)) \cong \mathbf{Z}[[x_1, \ldots, x_n]]$ on n variables. In particular, at least one of the λ-ring structures on the n-variable power series ring is topologically realizable.

Problem 8.9. *Classify the λ-ring structures on the power series ring on n variables that are topologically realizable.*

Since the K-theory of a topological space is a filtered λ-ring, it may be more natural to consider the above Problem for filtered λ-rings. For example, in $K(BSp(n))$ the variable x_i has filtration exactly $4i$. It is proved in [Yau (2004b)] that uncountably many of the isomorphism classes of filtered λ-ring structures on $\mathbf{Z}[[x_1, ..., x_n]]$ with x_i in filtration exactly $4i$ are topologically realizable. It is not known, however, if all such isomorphism classes are topologically realizable.

A variant of Problem 8.9 is to consider quotients of the n-variable power series ring, such as

$$K(\mathbf{HP}^\infty \times \mathbf{HP}^n) \cong \mathbf{Z}[[x, y]]/\langle y^{n+1}\rangle$$

with x and y both in filtration exactly 4. Here $\mathbf{HP}^n$ denotes the quaternionic projective n-space.

Next we concentrate on the one-variable truncated polynomial rings. In Theorem 6.13 it is proved that the truncated polynomial ring $\mathbf{Z}[x]/\langle x^n\rangle$ for $2 \leq n < \infty$ admits uncountably many isomorphism classes of filtered λ-ring structures. When $n = 2$ countably many of them are topologically realizable. Indeed, with the notation of Corollary 6.17, for $k \geq 1$ $K(S^{2k}) = \alpha((p^k))$ are mutually non-isomorphic filtered λ-ring structures on $\mathbf{Z}[x]/\langle x^2\rangle$. Here S^{2k} denotes the $2k$-sphere.

Notice that S^{2k} is a *torsion-free space*, by which we mean a space whose integral cohomology is $\mathbf{Z}$-torsion-free. Basic homotopy theory tells us that $K(S^{2k})$ $(k \geq 1)$ are the *only* filtered λ-ring structures on $\mathbf{Z}[x]/\langle x^2\rangle$ that are topologically realizable by torsion-free spaces.

For higher values of n, the situation is quite different. Indeed, for $3 \leq n \leq 6$, among the uncountably many isomorphism classes of filtered λ-ring structures on the truncated polynomial ring $\mathbf{Z}[x]/\langle x^n\rangle$, only *finitely many* of them are topologically realizable by torsion-free spaces. This is Corollary 1.13 in [Yau (2006)].

More explicit upper bounds can be obtained. For example, when $n = 3$, at most 64 isomorphism classes of filtered λ-ring structures on $\mathbf{Z}[x]/\langle x^3\rangle$ are topologically realizable by torsion-free spaces. In the notation of Theorem 6.21, these candidates for topological realization are $S((p^r), k)$ for $r \in \{1, 2, 4\}$ [Yau (2006)]. A corresponding upper bound when $n = 4$ is 61 [Yau (2006)].

Problem 8.10. *Prove or disprove: For each integer $n \geq 7$ only finitely many isomorphism classes of filtered λ-ring structures on $\mathbf{Z}[x]/\langle x^n \rangle$ are topologically realizable by torsion-free spaces.*

The reader is referred to the paper [Yau (2006)] for discussion about possible ways to approach Problem 8.10.

8.5 Mislin Genus

The next question is about the Mislin genus of a space. Given a 1-connected space X, it is possible to localize it rationally or with respect to a prime p [Sullivan (2005)]. The rationalization and the p-localization, where p is a prime, are denoted by $X_{(0)}$ and $X_{(p)}$, respectively.

Two 1-connected spaces X and Y are said to have the same *Mislin genus* if:

(1) $X_{(0)}$ and $Y_{(0)}$ are homotopy equivalent, and
(2) $X_{(p)}$ and $Y_{(p)}$ are homotopy equivalent for each prime p.

The Mislin genus of a 1-connected space X is the set of homotopy types of 1-connected spaces Y that have the same Mislin genus as X.

The Mislin genus of the classifying space of a Lie group is usually highly nontrivial. For example, the Mislin genus of the classifying space $BSU(2)$ of the special unitary group $SU(2)$ contains uncountably many distinct homotopy types [Rector (1971)]. More generally, if G is a 1-connected compact Lie group, then the Mislin genus of its classifying space BG also contains uncountably many distinct homotopy types [Møller (1992)]. In this case, Notbohm [Notbohm (1993)] showed that the Mislin genus of BG is classified by K-theory filtered λ-rings. In other words, any two distinct homotopy types X and Y in the Mislin genus of BG have non-isomorphic K-theory filtered λ-rings.

On the other hand, the Mislin genus of the quaternionic projective space $\mathbf{HP}^n$ ($2 \leq n < \infty$) has only finitely many distinct homotopy types [McGibbon (1982)]. For example, there are exactly four distinct homotopy types in the Mislin genus of $\mathbf{HP}^2$. In contrast to the cases of BG, it is proved in [Yau (2006)] that the Mislin genus of $\mathbf{HP}^2$ is *not* classified by K-theory filtered λ-rings. In other words, there exist two distinct homotopy types X and Y in the Mislin genus of $\mathbf{HP}^2$ such that $K(X)$ and $K(Y)$ are isomorphic as filtered λ-rings. This leads naturally to the following question.

Problem 8.11. *Prove or disprove: For $2 < n < \infty$, the Mislin genus of $\mathbf{HP}^n$ is not classified by K-theory filtered λ-rings.*

There is one curious feature of Notbohm's proof [Notbohm (1993)] of the K-theory classification of the Mislin genus of BG. Namely, it does *not* involve computing the λ-operations on the K-theory of the spaces in the Mislin genus of BG. It would be nice to better understand *how* the K-theory filtered λ-rings of the spaces in the Mislin genus of BG are different. This leads us to the following problem.

Problem 8.12. *Let G be a 1-connected compact Lie group, and let BG be its classifying space.*

(1) For each homotopy type X in the Mislin genus of BG, compute the K-theory filtered λ-ring of X.
(2) Show by purely algebraic methods that these filtered λ-rings are mutually non-isomorphic.

The first non-trivial case $BSU(2)$ is done in [Yau (2005)]. In this case, the Adams operations can be written in terms of the Rector invariants [Rector (1971)]. For other classifying spaces BG, one might need to develop analogs of the Rector invariants.

Hints to Selected Exercises

Chapter 1
(6). In the definition of P_n, set all the x_i to 0, and see what happens.
(7a). Use (1.1). This is Lemma 3.1.2 in [Atiyah (1989)].
(12). Use the free λ-ring U_2 on two generators from Exercise (11) instead of U.

Chapter 2
(1). First prove it for 1-dimensional elements. Then extend it to the general case by considering sums of 1-dimensional elements.
(6). The structure map $\lambda_t \colon R \to \Lambda(R)$ is an injective ring homomorphism. The $\mathbf{Q}$-algebra structure on R is given by an injective ring homomorphism $\varphi \colon \mathbf{Q} \to R$. Composing these two maps, we obtain an injective ring homomorphism $\lambda_t\varphi \colon \mathbf{Q} \to \Lambda(R)$, giving rise to a $\mathbf{Q}$-algebra structure on $\Lambda(R)$ that extends the one on R.

Chapter 3
(3). See Example 3.25 for the polynomial G_n.
(6). If $\{x_n\}$ is a Cauchy sequence in $\mathbf{Z}$ with respect to the p-adic metric, then the congruence class of x_n modulo p^m is constant for large n. Define a map $\pi_m \colon \hat{\mathbf{Z}}_p \to \mathbf{Z}/p^m$ by setting

$$\pi_m\left(\{x_n\}\right) = \text{congruence class of } x_n \text{ modulo } p^m \text{ for large } n.$$

Show that the maps π_m give the desired ring isomorphism.
(7c). Use Theorem 3.49. This problem is taken from p.54 in [Knutson (1973)].

Chapter 4
(1). Follow the procedure in Example 4.8.
(2). The additive inverse $\mathbf{b} = -\mathbf{a}$ is given by

$$b_n = \begin{cases} -1 & \text{if } n = 2^q \text{ for some } q \geq 0, \\ 0 & \text{otherwise.} \end{cases}$$

It suffices to prove this when R is $\mathbf{Z}$-torsion-free. First show that $b_p = 0$ whenever p is an odd prime. Use this to show that $b_n = 0$ when $n > 1$ is odd. Then show by induction that $b_{2^n m} = 0$ for $n \geq 1$ and $m > 1$ odd.

(5). It suffices to do this when R is $\mathbf{Z}$-torsion-free. Start with 1-dimensional elements. Then use the linearity of the pth power map modulo p.
(7). Use Wilkerson's Theorem 3.54.

Chapter 5

(1). See Example 5.19 and the discussion after the proof of Lemma 5.30.
(2). This is Proposition 2.3 in [Elliott (2006)]. To show that φ is injective, it suffices to show that the map

$$\phi\colon \bigotimes_{x\in X} \mathrm{Int}(\mathbf{Z}^{\{x\}}) \to \bigotimes_{x\in X} \mathbf{Q}[x]$$

induced by the inclusion maps is injective because the diagram

$$\begin{array}{ccc} \bigotimes_{x\in X} \mathbf{Q}[x] & \xrightarrow{\cong} & \mathbf{Q}[X] \\ \phi \uparrow & & \uparrow \\ \bigotimes_{x\in X} \mathrm{Int}(\mathbf{Z}^{\{x\}}) & \xrightarrow{\varphi} & \mathrm{Int}(\mathbf{Z}^X) \end{array}$$

is commutative. When X is finite, use the flatness of the free $\mathbf{Z}$-module $\mathrm{Int}(\mathbf{Z}^X)$ to show that ϕ is injective. Extend this to a general X using the exactness of direct limits. Surjectivity of φ follows from Theorem 5.28.
(3). This is Proposition 5.1 in [Elliott (2006)]. Use Corollary 5.11 (2).
(4). This is Proposition 5.4 in [Elliott (2006)]. Use Corollary 5.11 (2).
(5). This is part of Theorem 4.1 in [Elliott (2006)]. The polynomial $f/m \in \mathbf{Q}[x_1, \ldots, x_k]$ is an integer-valued polynomial, i.e., $f/m \in \mathrm{Int}(\mathbf{Z}^{\{x_1,\ldots,x_k\}})$. Now use Theorem 5.28.
(6). It remains to show distributivity and the associativity of the product. The first one is easy. For associativity first observe that it suffices to prove it for elements with only one non-zero entry.

Chapter 6

(1). Recycle the proof of Theorem 6.16. Since A is a $\mathbf{Q}$-algebra, every prime p is invertible in A, so the mod p condition on ψ^p is trivially true.
(2). Use the Cellular Approximation Theorem, which can be found in most books on algebraic topology, e.g., [Massey (1991); May (1999)].
(4). See Example 6.32 for the computation of the λ-operations on $S((c_p))$ (6.14).
(5). See Examples 6.23 and 6.24 for the cases $S((p), k)$ and $S((p^2), k)$, respectively.

Chapter 7

(1). Note that $(R, \{I_n\})$ is Hausdorff if and only if $\bigcap_n I_n = \{0\}$. Use this to prove the assertion about Hausdorff. For the other assertion, first observe that the map $p\colon R \to \varprojlim R/I_n$ is surjective if and only if the natural map $\alpha\colon \Lambda(R) \to \varprojlim \Lambda(R/I_n)$

is surjective. Then use the commutative diagram

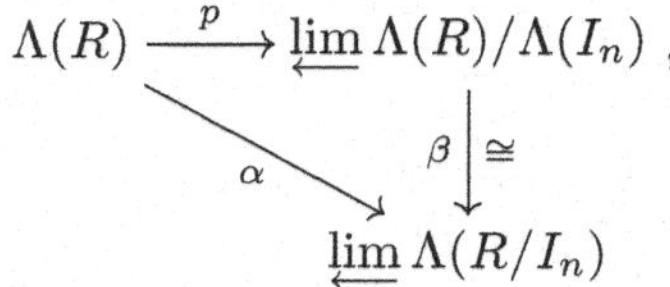

in which all the maps are the obvious ones and β is an isomorphism.

(4). This is almost exactly the same as Proposition 4.5 in [Lubin (1994)]. The coefficients of the power series $h(x) = cx + c_2x^2 + c_3x^3 + \cdots$ are constructed by induction, with the initial case being $h(x) \equiv cx \pmod{x^2}$. Suppose that the unique elements $c, c_2, \ldots, c_n$ have been constructed such that $h(g(x)) \equiv f(h(x)) \pmod{x^{n+1}}$. Then c_{n+1} exists such that $h(g(x)) \equiv f(h(x)) \pmod{x^{n+2}}$ if and only if $(a^{n+1} - a)$ is invertible in $\mathbf{k}$. This is the case by the hypothesis on a.

(5). The "only if" part is trivial. For the "if" part, consider the power series $\psi_2^p(\varphi(x))$ and $\varphi(\psi_1^p(x))$ for any prime p. They both have constant term 0 and have the same linear coefficient. Let $h(x)$ denote either one of them. Then $h(g(x)) = f(h(x))$ as power series, where $g(x) = \psi_1^q(x)$ and $f(x) = \psi_2^q(x)$. Now use the uniqueness part of Exercise (4) to conclude that $\psi_2^p(\varphi(x)) = \varphi(\psi_1^p(x))$ for every prime p. Since R is a subring of $\mathbf{Q}$, this implies that $\psi_2^p\varphi = \varphi\psi_1^p$. Now use Corollary 3.16.

List of Notations

The following notations are arranged in (roughly) alphabetical order, starting with arithmetic operations. For each notation, we provide the page number(s) where it is defined or first discussed and a brief description.

Notation	Page	Description
$+_\Lambda$	25	addition in $\Lambda(R)$
$+_W$	88	addition in $\mathbf{W}(R)$
$\times_\Lambda$	25	multiplication in $\Lambda(R)$
$\times_W$	88	multiplication in $\mathbf{W}(R)$
$\sum_\Lambda$	27	addition in $\Lambda(R)$
$\prod$	2	product
$\coprod$	140	disjoint union of sets
$\lceil s \rceil$	148	smallest integer $\geq s$
(i, j)	129	greatest common divisor of i and j
$[i, j]$	129	least common multiple of i and j
$\langle a, b \ldots \rangle$	135	ideal generated by $a, b, \ldots$
$\binom{n}{i}$	6	binomial coefficient
$\binom{x}{n}$	110	binomial symbol
$\binom{X}{I}$	119	generalized binomial symbol
$(-)_{ab}$	33	abelianization functor
Ab	32	category of abelian groups
BG	172	classifying space of the group G
$\mathbf{C}$	1	complex numbers
$\mathbf{CP}^n$	148	complex projective n-space
$\mathcal{C}(A, B)$	32	set of morphisms from A to B

$\lambda(R)$	142	isomorphism classes of (filtered) λ-ring structures on R
Λ	25	universal λ-ring functor
Λ^f	165	filtered version of Λ
L	67	ψ-isomorphism from $\Lambda(R \otimes \mathbf{Q})$ to $(R \otimes \mathbf{Q})^{\mathbf{N}}$
log	44	natural logarithm
$M(x, n)$	129	necklace polynomial
ν	70	counit of the comonad Ψ
$\nu_p(n)$	147	largest integer such that $p^{\nu_p(n)}$ divides n
$Nr(R)$	129	necklace ring of R
1_A	32	identity morphism of A
Ω_r	14	polynomial λ-ring $\mathbf{Z}[x_1, \ldots, x_r]$
Ω	15	inverse limit of Ω_r
Ω_r^γ	60	non-unital polynomial γ-ring $\mathbf{Z}[x_1, \ldots, x_r]^+$
Ω^γ	61	inverse limit of Ω_r^γ
$Ob(\mathcal{C})$	32	class of objects in a category $\mathcal{C}$
Op^γ	60	ring of natural operations on γ-rings
Op^λ	17	ring of natural operations on λ-rings
ψ^n	44	nth Adams operation
ψ^n	65	nth operation in a ψ-ring
ψ_{-t}	44	generating function for the ψ^n
Ψ	66	injective ψ-homomorphism from R to $R^{\mathbf{N}}$
Ψ	69	comonad whose coalgebras are ψ-rings
P_n	5	universal polynomial for the coefficient of t^n in $\prod(1 + x_i y_j t)$
$P_{n,m}$	4	universal polynomial for the coefficient of t^n in $\prod(1 + x_{i_1} \cdots x_{i_m} t)$
$\mathbf{Q}$	1	rational numbers
$\mathbf{Q}[X]$	116	ring of polynomials in X over $\mathbf{Q}$
Q_r	4	universal polynomial for the rth power sum
$Q_{n,m}$	98	universal polynomial for the coefficient of t^m in $\prod(1 + x_i^n t)$
$R^{\mathbf{N}}$	65	Cartesian product ring of R
$R(G)$	9	representation ring of a group G
$R[x_1, \ldots, x_n]$	1	polynomial ring in n variables over R
$R[[x_1, \ldots, x_n]]$	13	power series ring in n variables over R

Bibliography

Adams, J. F. (1960). On the non-existence of elements of Hopf invariant one, *Ann. Math.* **72**, pp. 20–104.

Adams, J. F. (1962). Vector fields on spheres, *Ann. Math.* **75**, pp. 603–632.

Adams, J. F. and Atiyah, M. F. (1966). K-theory and the Hopf invariant, *Quart. J. Math. Oxford* **17**, pp. 31–38.

Atiyah, M. F. (1989). *K-Theory* (Westview).

Atiyah, M. F. and Tall, D. O. (1969). Group representations, λ-rings and the J-homomorphism, *Topology* **8**, pp. 253–297.

Berthelot, P. *et al.* (eds.) (1971). *Théorie des Intersections et Théorème de Riemann-Roch, Lecture Notes in Math.*, Vol. 225 (Springer-Verlag, Berlin).

Berthelot, P. (1971). Generalites sur les λ-anneaux, in P. Berthelot *et al.* (eds.), *Théorie des Intersections et Théorème de Riemann-Roch, Lecture Notes in Math.*, Vol. 225 (Springer-Verlag, Berlin), pp. 297–364.

Borger, J. and de Smit, B. (2008). Galois theory and integral models of Λ-rings, *Bull. Lond. Math. Soc.* **40**, pp. 439–446.

Cahen, P. J. and Chabert, J. L. (1997). *Integer-Valued Polynomials, Math. Surveys and Monographs*, Vol. 48 (Amer. Math. Soc., Providence, RI).

Cameron, P. J. (1995). *Combinatorics: Topics, Techniques, Algorithms* (Cambridge Univ. Press, Cambridge, UK).

Clauwens, F. J. B. J. (1994). Commuting polynomials and λ-ring structures on $\mathbf{Z}[x]$, *J. Pure Appl. Alg.* **95**, pp. 261–269.

Dieudonné, J. (1957). On the Artin-Hasse exponential series, *Proc. Amer. Math. Soc.* **8**, pp. 210–214.

Dieudonné, J. (1973). *Introduction to the Theory of Formal Groups, Pure and Applied Math.*, Vol. 20 (Marcel Dekker, New York).

Dress, A. W. M. and Siebeneicher, C. (1988). The Burnside ring of profinite groups and the Witt vector construction, *Adv. Math.* **70**, pp. 87–132.

Dwork, B. (1958). Norm residue symbol in local number fields, *Abh. Math. Sem. Univ. Hamburg* **22**, pp. 180–190.

Elliott, J. (2006). Binomial rings, integer-valued polynomials, and λ-rings, *J. Pure Appl. Alg.* **207**, pp. 165–185.

Fulton, W. and Harris, J. (1991). *Representation Theory: A First Course, Graduate Texts in Math.*, Vol. 129 (Springer, New York).

Fulton, W. and Lang, S. (1985). *Riemann-Roch Algebra, Grund. der math. Wiss.*, Vol. 277 (Springer-Verlag, New York).

Graham, J. J. (1993). Generalised Witt vectors, *Adv. Math.* **99**, pp. 248–263.

Gromov, M. (1987). Hyperbolic groups. Essays in group theory, *Math. Sci. Res. Inst. Publ.* **8**, pp. 75–263.

Grothendieck, A. (1971). Classes de faisceaux et théorème de Riemann-Roch, in P. Berthelot *et al.* (eds.), *Théorie des Intersections et Théorème de Riemann-Roch*, *Lecture Notes in Math.*, Vol. 225 (Springer-Verlag, Berlin), pp. 20–77.

Guillot, P. (2007). The representation ring of a simply conneted Lie group as a λ-ring, *Comm. Alg.* **35**, pp. 875–883.

Hazewinkel, M. (1978). *Formal Groups and Applications*, *Pure and Applied Math.* Vol. 78 (Academic Press, New York-London).

Husemoller, D. (1994). *Fibre Bundles*, 3rd edn., *Graduate Texts in Math.*, Vol. 20 (Springer-Verlag, New York).

Jacobson, N. (1985). *Basic Algebra I* (W. H. Freeman).

James, G. and Liebeck, M. (2001). *Representations and Characters of Groups*, 2nd edn. (Cambridge Univ. Press, Cambridge, UK).

Knutson, D. (1973). *λ-Rings and the Representation Theory of the Symmetric Group*, *Lecture Notes in Math.*, Vol. 308 (Springer-Verlag, Berlin).

Knutson, D. (1975). Binomal λ-rings and a topological corollary, *Proc. Amer. Math. Soc.* **50**, pp. 83–84.

Kono, A. and Tamaki, D. (2002). *Generalized Cohomology*, *Translations of Math. Monographs*, Vol. 230 (Amer. Math. Soc., Providence, RI).

Lang, S. (2002). *Algebra*, 3rd edn. (Springer, New York).

Lenart, C. (1998). Formal group-theoretic generalizations of the necklace algebra, including a q-deformation, *J. Alg.* **199**, pp. 703–732.

Lubin, J. (1994). Nonarchimedean dynamical systems, *Comp. Math.* **94**, pp. 321–346.

Mac Lane, S. (1998). *Categories for the Working Mathematician*, 2nd edn., *Graduate Texts in Math.*, Vol. 5 (Springer-Verlag, New York).

Massey, W. S. (1991). *A Basic Course in Algebraic Topology*, *Graduate Texts in Math.*, Vol. 127 (Springer-Verlag, New York).

May, J. P. (1999). *A Concise Course in Algebraic Topology*, *Chicago Lectures in Math.* (Univ. Chicago Press, Chicago-London).

McGibbon, C. A. (1982). Self-maps of projective spaces, *Trans. Amer. Math. Soc.* **271**, pp. 325–346.

Metropolis, N. and Rota, G.-C. (1983). Witt vectors and the algebra of necklaces, *Adv. Math.* **50**, pp. 95–125.

Møller, J. M. (1992). The normalizer of the Weyl group, *Math. Ann.* **294**, pp. 59–80.

Munkres, J. R. (1975). *Topology: A First Course* (Prentice-Hall, New Jersey).

Notbohm, D. (1993). Maps between classifying spaces and applications, *J. Pure Appl. Alg.* **89**, pp. 273–294.

Osse, A. (1997). λ-structures and representation rings of compact connected Lie groups, *J. Pure Appl. Alg.* **121**, pp. 69–93.

Patras, F. (2003). Lambda-rings, in M. Hazewinkel (ed.), *Handbook of Algebra*, Vol. 3 (Elsevier, Amsterdam), pp. 961–986.

Rector, D. L. (1971). Loop structures on the homotopy type of S^3, in P. J. Hilton (ed.), *Symposium on Algebraic Topology*, *Lecture Notes in Math.*, Vol. 249 (Springer-Verlag, Berlin), pp. 99–105.

Royden, H. L. (1988). *Real Analysis*, 3rd edn. (Prentice-Hall, New Jersey).

Stanley, R. P. (1997). *Enumerative Combinatorics. Vol. 1* (Cambridge Univ. Press, Cambridge, UK).

Stanley, R. P. (2001). *Enumerative Combinatorics. Vol. 2* (Cambridge Univ. Press, Cambridge, UK).

Sullivan, D. P. (2005). *Geometric Topology: Localization, Periodicity, and Galois Symmetry, The 1970 MIT Notes* (Springer, Netherlands).

Weibel, C. A. (1994). *An Introduction to Homological Algebra, Cambridge Studies in Math.*, Vol. 38 (Cambridge Univ. Press, Cambridge, UK).

Wilkerson, C. (1982). Lambda-rings, binomial domains, and vector bundles over $\mathrm{CP}(\infty)$, *Comm. Algebra* **10**, pp. 311–328.

Yau, D. (2003a). Maps to spaces in the genus of infinite quaternionic projective space, *Progress in Math.* **215**, pp. 293–302.

Yau, D. (2003b). Unstable K-cohomology algebra is filtered λ-ring, *Int. J. Math. Math. Sci.* **2003**, pp. 593–605.

Yau, D. (2004a). Moduli space of filtered λ-ring structures over a filtered ring, *Int. J. Math. Math. Sci.* **2004**, pp. 2065–2084.

Yau, D. (2004b). On adic genus and lambda-rings, *Trans. Amer. Math. Soc.* **357**, pp. 1341–1348.

Yau, D. (2005). Loop structures on the homotopy type of S^3 revisited, *Michigan Math. J.* **53**, pp. 283–290.

Yau, D. (2006). On λ-rings and topological realization, *Int. J. Math. Math. Sci.* **2006**, Article ID 91267, pp. 1–21.

Index